Induction Heat Treatment of Steel

S. L. Semiatin and D. E. Stutz

Center for Metals Fabrication
Columbus Laboratories of
Battelle Memorial Institute

EPRI Project Management
I. Leslie Harry

American Society for Metals
Metals Park, Ohio 44073

Editorial and production coordination by
Carnes Publication Services, Inc.
Project managers: Craig W. Kirkpatrick and
Jeffrey A. Lachina

Library of Congress Catalog Card No.: 85-72048
ISBN: 0-87170-211-8
SAN 204-7586

To our families

Preface

Induction heat treatment of steel combines one of the oldest technologies, steel heat treating, with one of the most modern, induction heating. Heat treatment is the key to varying the properties of steels over a wide range, making them suitable for numerous applications. This is done by controlling the phase transformations which steels undergo in the solid state. Both time and temperature have an important influence on these transformations.

Traditionally, most heat treating operations have been carried out in fuel-fired furnaces. Such processing typically requires long heating times for completion. Through the application of induction, in which heating is accomplished by inducing electric currents within metal parts, times for many processes can be cut to minutes or even seconds. Induction heating is also attractive for applications involving surface and selective heat treating. In these cases, parts can be produced with a special blend of properties that cannot be achieved via other processing methods.

This book has been written for use by those who do not necessarily have backgrounds in metallurgy or electrical engineering. It is designed to be an introductory text or reference for undergraduate college students as well as practicing engineers. The discussion centers on the important aspects of steel heat treatment and the application of induction heating for conducting such processes. Following a brief introduction, Chapters 2, 3, and 4 deal with the fundamental principles involved in the heat treatment of steel and induction heating. In Chapters 5 and 6, the specific applications of induction heating for hardening and tempering, respectively, are described. Special considerations (e.g., post-heat-treatment properties) and uses of induction heat treatment are given in Chapters 7 and 8. The final two chapters deal with other heat treating processes for which induction may be applied and a discussion of the economics of induction heat treatment. Various appendices deal with microstructure development in steels, fundamental principles of electricity, and design of induction coils in more detail than is possible in the main body of this volume.

The impetus for writing this book came from several sources. First, and perhaps foremost, there is no other book that deals with both the metallurgical and electrical aspects of induction heat treatment. There are many books which deal with one or the other, but none which presents a balanced synthesis of the subject.

The present economic climate and the need for energy efficiency in industry have also spurred the undertaking of this project. With the volatility of both supply and price of all energy sources in recent years, the use of electricity for process heating applications in industry has seen significant increases. Electricity-based induction heating is already economically competitive with more conventional techniques and will become more so in the future. Thus, there exists a need for up-to-date information in order to apply induction heating efficiently for heat treatment.

The authors wish to express their gratitude to the organizations who made possible the writing and publication of this book. The bulk of the work was sponsored by the Electric Power Research Institute through its contract with Battelle Columbus Laboratories in establishing a Center for Metals Fabrication (CMF). The CMF conducts programs that promote the use of efficient electrotechnologies in industry. One of the major objectives of the Center is to encourage the efficient use of electricity for industrial process heating.

Thanks are also due to the publisher, the American Society for Metals. Having its roots in the area of steel heat treating, ASM has a long history of disseminating information to heat treaters. The authors appreciate the efforts taken by the ASM staff—in particular Mr. Timothy Gall—as well as by Ms. Laura Cahill, CMF Operations Coordinator, in the production of this book. The authors also are grateful to Craig Kirkpatrick and Jeffrey Lachina, Project Managers on the staff of Carnes Publication Services, Inc., for their professional skill in copy editing the manuscript.

A number of the authors' colleagues have also made substantial contributions through discussions and comments on the initial drafts of this volume. These include J. Cachat (TOCCO), T. Groeneveld (Battelle), P. Hassell (Hassell Associates), P. Miller (IPE Cheston), L. Moliterno (Ajax Magnethermic), G. Mucha (TOCCO), N. Ross (Ajax Magnethermic), R. Sommer (Ajax Magnethermic), H. Udall (Thermatool), B. Urband (Tubulars Unlimited), D. Williams (Battelle), and S. Zinn (Ferrotherm, Inc.). T. Groeneveld, P. Miller, G. Mucha, N. Ross, C. Tudbury, and S. Zinn also supplied some of the illustrations used herein. The efforts of C. Sullivan, B. Pierpoint, and R. Underwood in preparing the manuscript are also appreciated.

Lastly, the authors offer special thanks to Mr. Tom Byrer, Director of the Center for Metals Fabrication, and Mr. Les Harry, Program Manager at the Electric Power Research Institute. Without their support and encouragement, the writing of this book would not have been possible.

Lee Semiatin
Dave Stutz
January, 1985

Contents

Chapter 1

Introduction

". . . [When the steel pieces] are very hot and almost of a white color because of the heat, in order that the heat may be quickly quenched, they are suddenly thrown into a current of water that is as cold as possible . . . In this way, the steel is made so hard that it surpasses almost every other hard thing. . . "

V. Biringuccio, *Pirotechnia,* 1540.

Induction heat treatment of steel combines what is surely one of the oldest of all technologies, steel heat treatment, with one of the more recent, induction heating. Hardening of steel through heat treatment had its origins during the Iron Age (circa 1000 B.C.), but it was not until the latter part of the 19th century and the first part of the 20th century that this process (and the metallurgical changes that are responsible for it) came to be fully understood. Electromagnetic induction, or simply "induction," was first discovered in the experiments of Michael Faraday around 1830. Faraday noted that an electric current could be induced in an electrical conductor by suitably coupling the conductor with a coil carrying an alternating current. Such a current can give rise to heating, and thus was formed the basis for techniques such as induction melting and induction hardening. Induction melting processes were probably the first to be exploited commercially. Later, during the 1930's, induction methods for hardening razor blades and the bearing surfaces of crankshafts were introduced. Today, steel products are being induction heat treated in rapidly increasing numbers.

Steels are the most common of all engineering materials. In 1981, approximately 100 million tons of steel were made in a multitude of product forms. An understanding of why so much steel is used requires consideration of both the production characteristics and the properties of steel products. Manufacture of the various forms of steel often starts with reduction of the raw material, iron ore, to produce pig iron. Pig iron is a high-carbon iron alloy which is further refined and alloyed to make different grades of steel. The process of ore reduction is relatively easy from an energy standpoint, a circumstance which is akin to the rapidity with which many steels rust. The process of rusting, or oxidation, is merely the reverse of the chemical reaction involved in ore reduction.

The wide variety of uses for which steel is appropriate provides a second major reason for the importance of steel as an engineering material. Often we think of steel as being a "strong" material—i.e., it can support large loads in service without deforming—but if strength were the only positive attribute of these alloys they would not be used as often as they are. Unlike many materials, steel can be alloyed and processed to obtain an amazing range of properties which can be tailored to the needs of the final application (Fig. 1.1). These properties include not only strength but also toughness (the ability to resist brittle fracture) and resistance to fatigue (cyclic loading), creep (deformation at high temperatures under load), corrosion, and abrasive wear.

For a given alloy composition, the primary methods by which properties are altered or controlled may be classified under the general headings of mechanical processes and thermal processes. As the name implies, mechanical processes comprise those means involving the use of plastic (permanent) deformation and include operations such as rolling, forging, and extrusion. Besides imparting a given shape to a piece of steel, these techniques may be employed to increase strength or develop a microscopic structure ("microstructure") which gives rise to attractive properties. Thermal processes, on the other hand, involve no shaping, but rely solely on microstructural changes during heat treatment to modify properties. Sometimes, a thermal process is used in conjunction with a mechanical, or deformation, process in what is commonly known as thermomechanical treatment

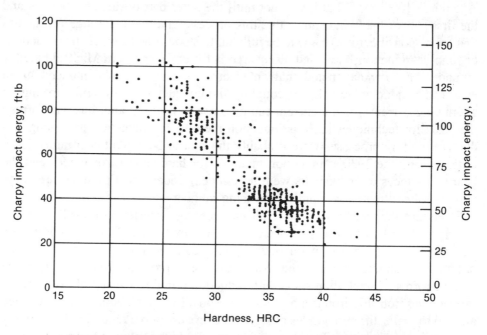

Fig. 1.1.   Relationship between room-temperature brittle fracture resistance (in terms of Charpy impact energy) and hardness for a variety of quenched-and-tempered plate steels containing 0.30 to 0.40% carbon (Ref 1)

(TMT) or thermomechanical processing (TMP). In this book, only the application of heat treatment to steel will be described. As an introduction, some of the common heat treating processes used for steel will be briefly discussed. Also, the heating methods by which these operations are carried out will be summarized.

HEAT TREATING OPERATIONS FOR STEEL

Although there are many different types of heat treating operations for steel, we shall concentrate on only a rather limited number of them. Most of our attention will be focused on the treatments known as hardening and tempering. Where appropriate, other important techniques, such as recrystallization annealing and spheroidization, will be discussed.

As its name implies, hardening is used to increase the strength of steel. Usually, this involves a heat treatment in which a "hard" microstructure known as martensite is produced. How this is done will be discussed in more detail in Chapter 2. In brief, conventional hardening processes involve raising the temperature of the steel above a certain critical temperature, holding it at that temperature for a fixed amount of time, and then rapidly cooling (i.e., quenching) it to room temperature (and sometimes to even lower temperatures). In some instances, the steel is air cooled. In these cases, the operation is known as "normalizing."

In the hardened condition, the steel is strong but relatively brittle. This is because of the general nature of the quenched microstructure and of the development of internal, residual stresses set up as a result of nonuniform cooling. To overcome this situation, the heat treating operation known as "tempering" is subsequently performed. During tempering, residual stresses are relieved and the microstructure is modified so that a much tougher, albeit sometimes slightly weaker, finished product is obtained. This is done at temperatures substantially lower than those used for hardening, but the time at temperature can be much longer. Steels which undergo both of the above processes are known as "quenched-and-tempered" steels.

The other heat treatment operations mentioned above (recrystallization annealing and spheroidization) are generally used to soften the steel and thus enhance the ability of the material to be deformed during further processing. Recrystallization annealing is a high-temperature heat treatment which is employed to eliminate the effects of prior mechanical working by producing new deformation-free metal grains in the metal. The new recrystallized microstructure has generally lower resistance to deformation during working. Common applications of this heat treatment include (*a*) hot rolled or press forged steel ingots and (*b*) cold rolled steel sheet. In the former application, the recrystallization anneal serves to replace the cast ingot microstructure, which is not very workable, with a recrystallized one. Recrystallization annealing of cold rolled sheet is employed to impart the material properties required for deep drawing of the steel into such products as auto-body and appliance panels.

The other important operation used to soften steel for further working is spheroidization annealing. In this process, the steel is heated to a moderate temperature in order to change the shape of the iron carbide (cementite) particles which are produced in the steel microstructure upon cooling from higher-temperature processing. During spheroidization, the carbide particles agglomerate and become spheroidal. In this condition, the steel is as soft and ductile as it can be, making it suitable for such demanding forming operations as cold forging.

HEAT TREATING METHODS

Heat treating methods may be broadly classified as either indirect or direct. In indirect methods, heat is produced in a furnace through combustion of a fuel or by conversion of electrical energy into heat by passing a current through resistance heating elements. This energy is then transferred to the material to be heated (known as the workpiece) by radiation, convection, or conduction. In direct methods, heat is supplied directly to the workpiece without the use of a furnace. Examples of such methods include flame heating, induction heating, and direct-contact resistance heating.

For the indirect methods of heat treatment of steel, gas-fired furnaces and electric-resistance furnaces are probably the most common types of equipment. The gas-fired furnace, the workhorse of the steel heat treating industry, provides heat through combustion of natural gas. This heat is transferred to the steel in the furnace by convection (often by forced convection obtained by use of blower fans in the furnace) as well as by radiation and conduction of heat from the furnace lining of high-temperature refractory material. In the electric furnace, heat energy passes into the workpiece primarily by radiation from the electric-resistance heating elements and secondarily by conduction and convection. A third type of indirect heating equipment is the molten salt or metal bath. Typically used for small production lots, the salt or metal medium is itself heated by electric-resistance or gas-fired heaters. Once the bath is liquid, the material to be heat treated is placed in it and heated by conduction. In gas-fired and electric furnaces, the hardening and tempering operations for steel typically last between 30 minutes and several hours depending on section thickness. This time may be considerably shortened by use of salt or metal baths.

Heat treating time may be further reduced by using direct methods such as induction. With this method, electric currents are induced in the steel by surrounding it with a coil (also known as an inductor) of a configuration similar to that of the steel workpiece. An alternating current (ac) is passed through the coil. Associated with the current is a magnetic field whose magnitude varies directly with the current. This magnetic field penetrates into the workpiece and induces alternating currents (known as eddy currents) as the magnetic field changes. Because the strength of the magnetic field decreases with distance from the workpiece surface, the induced eddy currents vary in magnitude with position as well. Hence, the rate at which electrical energy is converted into heat, and thus the temperature, varies with position. This effect forms the basis for surface heat treatment of steel. Induction may also be used for through-heating and for through heat treatment. This is done by selection of equipment which provides a relatively low-frequency alternating current to the inductor. By this means, the "penetration depth" of the eddy currents is increased. In addition, the power input to the workpiece is made sufficiently low so that the heat is generated slowly at the surface of the workpiece and therefore has time to be transferred to the center, thereby decreasing the thermal gradient.

Each of the above heat treating processes has both advantages and disadvantages. For example, furnace treatments are well suited to processing of large numbers of parts (often irregular in shape) in batch-type operations. Also, because heating is done slowly, the over-all heating time can be varied within relatively wide limits without greatly affecting the properties of the final product. On the negative side, the long heating times required in furnace treatments generally give rise to large amounts of surface scaling unless the furnace atmosphere is controlled. Furthermore, furnaces — particularly those of the gas-fired type — tend to be relatively inefficient from an energy standpoint.

Induction heat treatment precludes many of the problems associated with furnace methods. Among its advantages is the rapid heating that can be achieved. For this reason, induction heat treatment is particularly well-suited to high-volume continuous heat treatment operations. With the advent of microprocessor technology, the controls necessary for such techniques have become readily available. The rate of heating is limited only by the power rating of the ac power supply and the need for through-heating rather than surface heating. Because heating times are usually short, surface problems such as scaling and decarburization and the need for protective atmospheres can often be avoided. In addition, induction tends to be energy-efficient. With proper coil design and equipment selection, more than 80% of the electrical energy can be converted into heat for treatment of the workpiece. Such efficiencies are impossible with gas-fired furnaces, in which a large proportion of the consumed energy is lost with the hot gases leaving the furnaces. Induction heating is also free of pollution, a consideration to which users of gas-fired furnaces should pay careful attention. The above attributes have resulted in the application of induction heating to a variety of heat treating processes. An estimate of the extent of its utilization is presented in Fig. 1.2.

Among the disadvantages of induction are those related to coil design and equipment selection, both of which must be tailored to the particular part to be heat treated and the temperature at which the heat treatment is to be carried out. In the automotive and oil-drilling equipment industries, production rates are high and the induction heat treating method finds wide application. In situations where only a few parts of a given design are to be made, induction heat treatment is usually not economically feasible.

SCOPE OF THE BOOK

The chapters that follow will center on the important aspects of steel heat treatment and the application of the induction heating method for this purpose. Chapters 2, 3, and 4 will deal with the fundamental principles involved in the heat treatment of steel and induction heating. In Chapters 5 and 6, the specific appli-

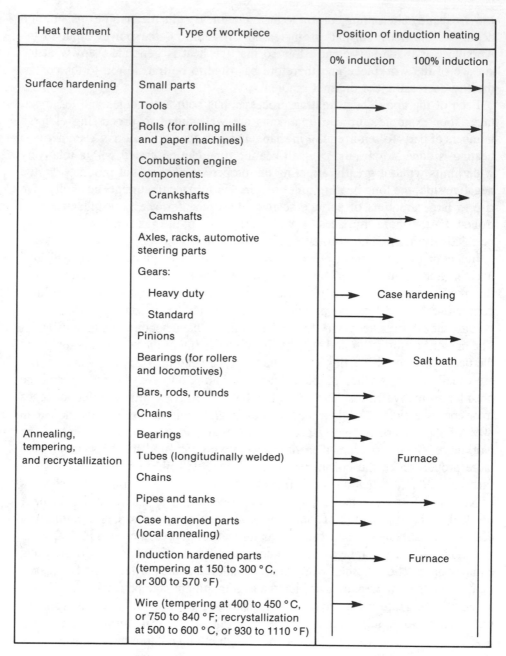

Heat treatment	Type of workpiece	Position of induction heating
Surface hardening	Small parts	
	Tools	
	Rolls (for rolling mills and paper machines)	
	Combustion engine components:	
	Crankshafts	
	Camshafts	
	Axles, racks, automotive steering parts	
	Gears:	
	Heavy duty	Case hardening
	Standard	
	Pinions	
	Bearings (for rollers and locomotives)	Salt bath
Annealing, tempering, and recrystallization	Bars, rods, rounds	
	Chains	
	Bearings	
	Tubes (longitudinally welded)	Furnace
	Chains	
	Pipes and tanks	
	Case hardened parts (local annealing)	
	Induction hardened parts (tempering at 150 to 300 °C, or 300 to 570 °F)	Furnace
	Wire (tempering at 400 to 450 °C, or 750 to 840 °F; recrystallization at 500 to 600 °C, or 930 to 1110 °F)	

Fig. 1.2.   Estimated usage of induction relative to all other heat treating processes (Ref 2)

cations of induction heating to hardening and tempering, respectively, are described. Special considerations and specific uses and case histories of induction heat treatment are given in Chapters 7 and 8. The final chapters deal with other heat treating processes to which induction may be applied (Chapter 9) and a discussion of the economics of induction heat treatment (Chapter 10).

Chapter 2

Basic Principles of Steel Heat Treatment

As was mentioned in Chapter 1, one of the main reasons why steel finds such wide application is the fact that its properties can be varied extensively by changes in chemical composition and heat treatment. Steels can be made relatively soft so that they can be shaped and subsequently hardened to various levels to provide strength for severe service applications. In this chapter, the basic principles of hardening of steel are reviewed. This discussion will be short and will describe only those aspects of practical significance which must be understood before heat treating systems based on induction can be designed. For a more complete description, the reader is referred to one of the standard books on steel heat treatment (Ref 3 to 7).

GENERAL BACKGROUND ON STEEL HEAT TREATMENT

The chemical compositions of commercial steels may include as few as two primary elements — iron and carbon — or, for steels designed for special service, as many as five or more elements. It might seem impossible to understand, let alone control, the heat treatment of such different forms of the same product, especially in light of the fact that there are thousands of different types of steel, but it must be realized that many steels can be grouped together according to the kinds of microstructures which can be developed in them through heat treatment. We have said previously that our attention will be focused on the quenched-and-tempered grades. Although different in composition, all steels under this classification respond similarly to heat treatment.

Hardening of quenched-and-tempered steels relies heavily on the fact that iron and many of its alloys undergo what is known as an allotropic transformation. This means that, depending on temperature, the metal in its solid state assumes different phases or crystal structures. For iron, the allotropic forms are two: a body-centered-cubic (bcc) structure found at low to moderately high temperatures *and* at very high temperatures; and a face-centered-cubic (fcc) structure found at

7

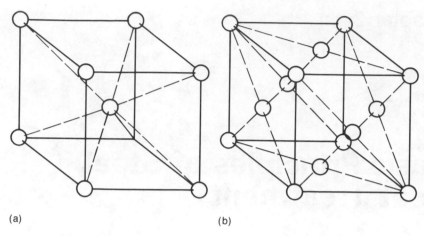

    (a)   (b)

Fig. 2.1. Crystal lattices found in iron and steel: (a) body-centered cubic (bcc) and (b) face-centered cubic (fcc)

moderately high temperatures (Fig. 2.1). These phases are referred to by the Greek letters alpha (α) and delta (δ) for the two bcc structures and gamma (γ) for the fcc structure. Under equilibrium conditions, the transformations occur at temperatures of 910 °C (1670 °F) for the α-to-γ change and 1400 °C (2550 °F) for the γ-to-δ change.*

In the above discussion, we have used the thermodynamics term "equilibrium conditions," which means the phase or condition of the metal that possesses the lowest amount of internal energy. This energy is tied up in the vibration and stacking arrangement of the atoms that comprise the regular array that characterizes crystalline materials. In pure iron, equilibrium is achieved very quickly at a given temperature, and one can expect to obtain the α, γ, and δ forms of iron by merely changing the temperature.

The heat treatment of steel relies on the fact that equilibrium is *difficult* to achieve in alloys of iron and other elements, principally carbon. In this way, "metastable," nonequilibrium microstructures (structures that are stable for relatively long times at room temperature and relatively modest higher temperatures) may be produced. The way in which these microstructures are developed is most readily understood by reference to graphical representations known as phase diagrams, which show the equilibrium phases achievable in alloys as a function of temperature and composition. The simplest phase diagram of importance in steel technology is the "binary" iron-carbon diagram presented in Fig. 2.2.

The iron-carbon phase diagram shows that, besides the liquid phase at high temperatures, the solid phases in iron alloys with varying amounts of carbon may also assume the α, γ, or δ crystal structure. In alloys with very small amounts of carbon, the alloy may be all α, all γ, or all δ at various temperatures. With

---

*Conversion factors for temperatures and other quantities frequently used in induction heat treatment are tabulated in Appendix A.

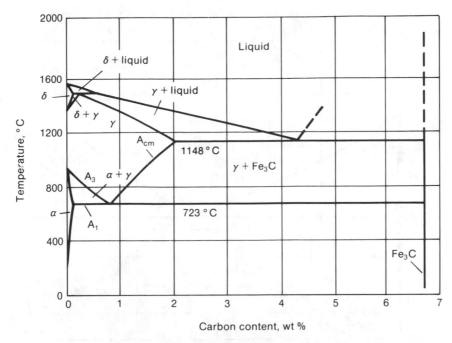

Although Fe₃C (cementite) is only a metastable compound, very long times are required for its decomposition into ferrite and graphite. For practical purposes, therefore, this diagram is usually referred to as the equilibrium phase diagram for the Fe-C system.

Fig. 2.2. The iron-carbon, binary phase diagram (Ref 8)

increasing carbon content, two phases are present except at certain carbon contents and temperatures at which γ-iron, or austenite, is found. Thus, we say that the "solubility" of carbon in α-iron (α-ferrite, or simply "ferrite") and in δ-iron (δ-ferrite) is low. This usage of the term "solubility" in connection with solid-state compounds is exactly analogous to its usage with regard to liquid solutions studied in chemistry. As we shall see later, the phenomenon of precipitation in liquid solutions has an analogy in solid-state reactions as well.

Returning to the phase diagram shown in Fig. 2.2, it can be seen that, depending on temperature and carbon content, various combinations of the α, γ, δ, and liquid phases, as well as phases referred to as iron carbide (or simply "carbide") and graphite (pure carbon), can be obtained. Iron carbide is an intermetallic compound of the exact (or "stoichiometric") composition Fe₃C. Given enough time (thousands of years), it would decompose into α-iron (or, at higher temperatures, γ-iron) and graphite. Thus, under true equilibrium conditions, the vertical line representing Fe₃C in Fig. 2.2 would not be present, and the carbide phase would be replaced by graphite everywhere it appears.

We are now in a position to define precisely what steels are. This definition is based on phase diagrams such as the one shown in Fig. 2.2. Steels are taken to be those iron alloys in which a totally γ-iron, or austenitic, structure can be developed by heating to some specific temperature. In some steels, alloying

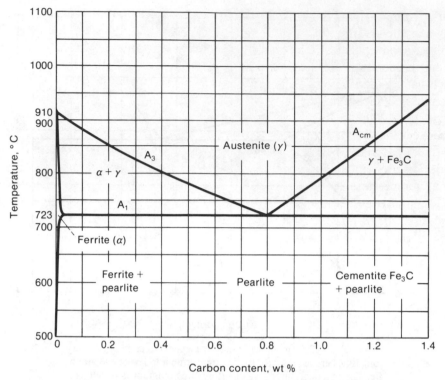

Fig. 2.3. The lower left-hand portion of the iron-carbon phase diagram

elements are added to make the austenitic phase stable even at (and below) room temperature. Examples of these are the austenitic stainless steels. In others, such as quenched-and-tempered steels, alloying additions are controlled in order to make the austenite phase stable only at high temperatures. Heating into the γ-phase field and subsequent decomposition of the austenite phase as the temperature is lowered constitute the process that serves as the basis for hardening. Note that at carbon contents above about 2 wt % it is impossible to obtain a totally austenitic structure in iron-carbon alloys at any temperature. These compositions are known as cast irons.

The process of decomposition, or transformation, of austenite as the temperature is lowered is best understood by referring again to the iron-carbon phase diagram. For this purpose, a slightly magnified version of the iron-carbon diagram at the low-carbon end is presented in Fig. 2.3. It can be seen that, *above* and *below* 0.8% C, austenite in a steel will partially decompose into a mixture of ferrite and austenite (%C < 0.8) or Fe_3C and austenite (%C > 0.8)* as the temperature is lowered below the critical temperatures which define the lower

---

*The ferrite and Fe_3C thus formed between the upper and lower critical temperatures are termed *proeutectoid* constituents since they form above the eutectoid temperature (whose significance is discussed below).

limits of the γ-phase field. These critical temperatures are delineated by lines known as the A_3 line (%C < 0.8) and the A_{cm} line (%C > 0.8).*

When another critical temperature, denoted by A_1 and known as the "lower critical temperature" in steel terminology, is reached, all remaining austenite which has not decomposed will do so. Note that for a 0.8% carbon steel all the austenite decomposes at the A_1 temperature. This reaction is called a "eutectoid reaction," and the 0.8% carbon content is called the "eutectoid composition." Moreover, the A_1 temperature is frequently referred to as the "eutectoid temperature," because all austenite in steels of other carbon contents will undergo a similar reaction at this temperature as well. The phase diagram shows that below the A_1 temperature austenite will turn into ferrite and Fe_3C under equilibrium conditions. Because it involves the lowest internal energy, the equilibrium ferrite-Fe_3C mixture, or morphology, will consist of a number of Fe_3C spheroids in a "matrix" of ferrite. This can only be achieved, however, if the cooling rate is very low and if the unstable austenite is held just below the A_1 temperature for a very long time (on the order of many hours). With somewhat higher cooling rates, or shorter holding times at lower temperatures, the austenite will decompose into a harder ferrite-Fe_3C mixture which has a lamellar appearance; alternating lamellae consist of the ferrite and Fe_3C constituents. Because of its resemblance to mother-of-pearl, this microstructure, or morphology, is called "pearlite."† With yet higher cooling rates, or shorter holding times at even lower temperatures, a much finer microstructure of ferrite and Fe_3C, known as "bainite," is obtained. Finally, if the austenite is cooled very rapidly to a yet lower temperature denoted by M_s, the carbon dissolved in the γ phase will not have time to diffuse and bring about the large-scale atomic rearrangements necessary to produce ferrite and Fe_3C. In this case, the fcc austenite phase shears on the atomic planes to form a less symmetrical body-centered-tetragonal (bct) crystal structure in which the carbon is still dissolved. This microstructure is known as "martensite" and is very hard; the M_s temperature is the temperature at which martensite *starts* to form.

Once all the austenite has decomposed in a carbon (or other hardened) steel, the steel is usually further heat treated after reaching room temperature or after being quenched to a relatively low temperature such as 100 °C (210 °F). This heat treatment, called "tempering," usually occurs between 370 and 595 °C (700 and 1100 °F) and modifies the carbide shape in pearlitic and bainitic microstructures or results in the precipitation of carbon from solution in the martensitic microstructure to produce carbides. The microstructures produced thusly are known as tempered pearlite, tempered bainite, and tempered martensite, and the properties of steels with these microstructures are usually quite different from those in the untempered condition.

---

*It should be kept in mind that the A_3 and A_{cm} (and subsequently to be discussed A_1) temperatures apply only to equilibrium conditions. During all but the slowest heating and cooling cycles, the corresponding transformation temperatures are either higher (Ac_3, Ac_{cm}, and Ac_1 for heating) or lower (Ar_3, Ar_{cm}, and Ar_1 for cooling).

†Pearlite and other common microstructures developed during heat treatment of steels are illustrated in Appendix B.

Table 2.1.   Compositions of common AISI/SAE steels

| AISI/SAE number(a) | Composition(b), % | | | |
	Ni	Mo	Cr	Others
10xx........	...	...	...	...
11xx........	...	...	...	...
13xx........	...	...	...	1.5 to 2.0 Mn
23xx........	3.25 to 3.75	...	...	...
25xx........	4.75 to 5.25	...	...	...
31xx........	1.10 to 1.40	...	0.55 to 0.90	...
33xx........	3.25 to 3.75	...	1.40 to 1.75	...
40xx........	...	0.20 to 0.30	...	...
41xx........	...	0.08 to 0.25	0.40 to 1.20	...
43xx........	1.65 to 2.00	0.20 to 0.30	0.40 to 0.90	...
46xx........	1.40 to 2.00	0.15 to 0.30	...	...
48xx........	3.25 to 3.75	0.20 to 0.30	...	...
51xx........	...	...	0.70 to 1.20	...
61xx........	...	...	0.70 to 1.10	0.10 V
81xx........	0.20 to 0.40	0.08 to 0.15	0.30 to 0.55	...
86xx........	0.30 to 0.70	0.08 to 0.25	0.40 to 0.85	...
87xx........	0.40 to 0.70	0.20 to 0.30	0.40 to 0.60	...
92xx........	...	...	...	1.80 to 2.20 Si

(a) The xx indicates carbon content in hundredths of a percent (0.xx% C). (b) The 10xx grades are plain carbon steels. The 11xx grades are plain carbon steels, resulfurized for machinability. All steels contain at least 0.25 to 1.00% Mn in order to form MnS and thus enhance hot workability. Many steels contain certain amounts of aluminum and silicon which serve as deoxidizing agents during melting and casting. In many steels of the 13xx series and above, composition limits for particular grades differ slightly from those for the corresponding AISI/SAE grades for more precise control and prediction of hardenability; these steels are known as H-band steels and are designated by 13xxH, 23xxH, etc.

The iron-carbon steels just discussed are often called plain carbon steels because carbon is the only important alloying element.* These steels are also often designated according to American Iron and Steel Institute (AISI) standards by 10xx, where xx denotes the carbon content in hundredths of a percent. There are a large number of other steels whose heat treatment often involves hardening and tempering processes very similar to those used for plain carbon steels. The chemical composition ranges for some of the more common AISI grades are given in Table 2.1. The most common alloying elements (other than carbon) in these steels are manganese, nickel, chromium, and molybdenum. Besides imparting improved strength and other properties, these alloying elements are very useful in developing martensitic microstructures during hardening. More will be said about this later.

In addition to the AISI grades listed in Table 2.1, there are several other very specialized grades of quenched-and-tempered steels. These alloys include mar-

---

*Commercial plain carbon steels also contain intentional additions of manganese, silicon, and undesirable residual (or trace) elements such as phosphorus and sulfur.

tensitic stainless steels (which combine strength and corrosion resistance), high speed and oil-hardening tool steels (for metalcutting and metalworking applications), and chromium-, tungsten-, and molybdenum-base hot work die steels (for hot working applications).

In the next several sections, each of the steps in the heat treatment of quenched-and-tempered steels will be reviewed in more detail.

AUSTENITIZING

In the previous section, the initial step in the hardening of steel was shown to be the formation of γ-iron, or austenite. This process is called "austenitizing" and depends critically upon temperature and time. For plain carbon steels, the critical temperatures are delineated in the phase diagram in Fig. 2.2, which shows that steel of the eutectoid composition (0.8% C) is most readily austenitized—i.e., can be austenitized at the lowest temperature.

When other elements such as chromium, manganese, and molybdenum are added to steels, the boundaries of the γ-phase field change to a degree that depends on the amounts of the alloying additions and on the carbon content. Figures 2.4(a), (b), and (c) show this effect for varying amounts of chromium,

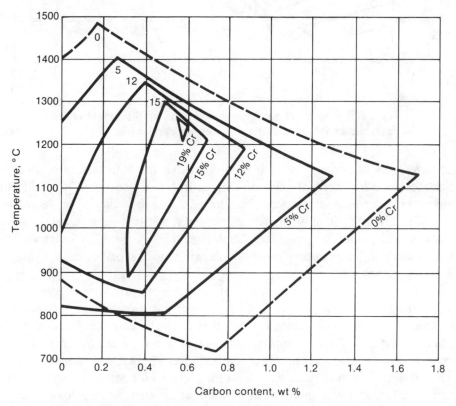

Fig. 2.4(a). Effect of chromium additions on the extent of the austenite phase field in the iron-carbon system (Ref 9)

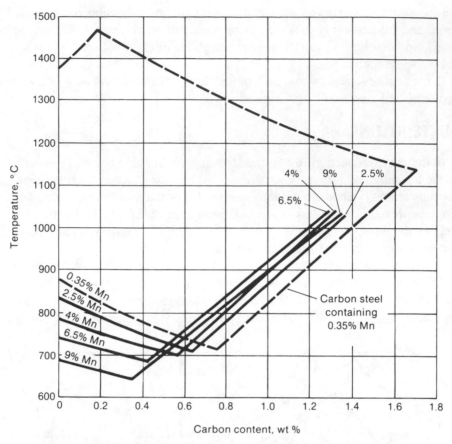

Fig. 2.4(b).   Effect of manganese additions on the extent of the austenite phase field in the iron-carbon system (Ref 9)

manganese, and titanium. Here, it is obvious that the eutectoid carbon content decreases as alloy content increases in both cases. However, the eutectoid temperature increases in one case but decreases in the other. The decrease in eutectoid carbon content with increasing alloy addition is a general one for all alloying elements of commercial importance (Fig. 2.5a). The exact variation of the eutectoid temperature, however, depends on the particular alloying element (Fig. 2.5b).

The time required to form a totally austenitic microstructure depends on the austenitizing temperature selected and on the starting microstructure. In all cases, the speed with which austenite is formed is controlled by the speed of carbon diffusion, a process which can be accelerated a great deal by an increase in temperature. For example, in a plain carbon steel of eutectoid composition with an initial microstructure of pearlite, the time required for complete austenitization can be decreased from approximately 400 s at an austenitizing temperature of 730 °C (1345 °F) to about 30 s at an austenitizing temperature of 750 °C (1380 °F), as shown in Fig. 2.6. Each of these times can be increased considerably

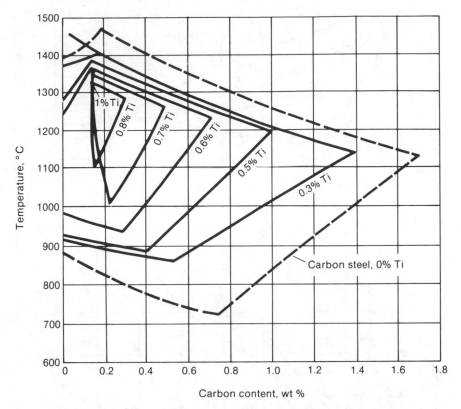

Fig. 2.4(c).   Effect of titanium additions of the extent of the austenite phase field in the iron-carbon system (Ref 9)

by starting with a spheroidized microstructure containing large carbide particles, because in such a microstructure the diffusion distance for carbon, which must be transported from the carbon-rich carbide phase, is considerably greater than the diffusion distance in a pearlitic structure with thin lamellae of ferrite and carbide. Conversely, the finer bainitic and martensitic microstructures tend to be re-austenitized more readily than pearlite.

Another example of a time-temperature relationship for reaustenitization is shown in Fig. 2.7 — in this case for a 0.2%C-10%Cr steel with an initial micro-structure consisting of a combination of ferrite (produced by prior cooling from the A_3 temperature to the A_1 temperature for this alloy) and a bainite-like structure containing complex iron-chromium carbides. In this figure, there are lines which mark the start (S) and completion or finish (F) of the austenite reversion process. At temperatures just above the A_3 temperature, this reaction takes approximately 100 s. The time can be shortened considerably (to only a few seconds) by increasing the temperature by 50 to 100 °C (90 to 180 °F). This effect explains why controlled, rapid heating to high austenitizing temperatures, such as is possible with induction heating, can minimize the required holding time for austenitizing prior to hardening by quenching.

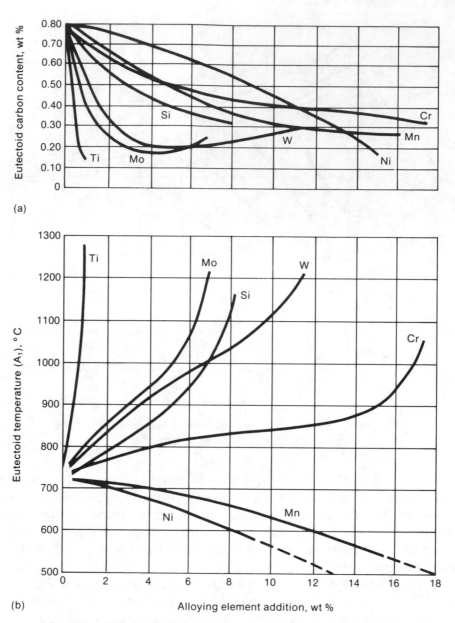

Fig. 2.5. Effects of alloying element additions on (a) eutectoid carbon content and (b) eutectoid temperature (Ref 10)

From data such as those just presented, *minimum* temperatures and times for austenitizing may be estimated. For a given austenitizing temperature, the time may be longer than the minimum estimated time. However, if the time at temperature is increased too much, the austenite grain size may become excessively large. The size of these grains (individual crystals of a given orientation) determines to a large extent the hardness and toughness of the final quenched-and-

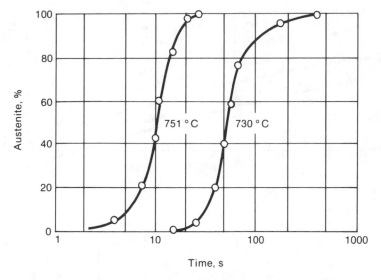

Fig. 2.6. Effect of austenitizing temperature on the rate of austenite formation from pearlite in a eutectoid steel (Ref 11)

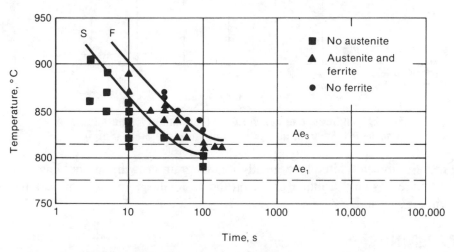

The letters "S" and "F" denote, respectively, the start and finish of the reaction as determined metallographically.

Fig. 2.7. Time-temperature-transformation curves for austenitization of a 10Cr-0.2C carbon steel with an initial microstructure of ferrite with fine fibrous and particulate iron-chromium carbides ($M_{23}C_6$) (Ref 12)

tempered steel. Both of these properties are improved by decreasing the austenite grain size. The problem of grain growth is of particular concern in "lean" steels (alloys with relatively small amounts of additional elements), especially for austenitizing temperatures around 1000 to 1050 °C (1830 to 1920 °F). For the more highly alloyed steels (such as hot work die steels), the presence of the alloying

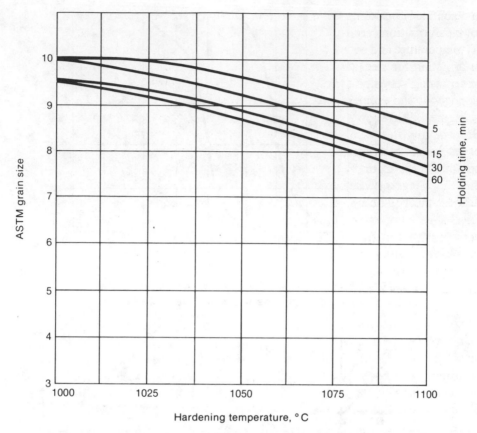

Fig. 2.8.   Influence of austenitizing (hardening) temperature and time on ASTM grain size in H13 hot work die steel (Ref 5)

elements in the austenite phase greatly retards grain growth, and holding time at temperature has only a minor effect on the austenite grain size, an example of which is given in Fig. 2.8.*

HARDENING

Following austenitizing, steel can be hardened to various degrees by controlling the cooling rate and the temperature at which the steel is held below the A_1 temperature for a given period of time prior to further cooling. The hardening reaction and the resulting microstructures, or transformation products, can be determined by reference to graphs known as isothermal transformation or IT diagrams (also referred to as time-temperature-transformation or TTT diagrams)

---

*The ASTM grain size number in this figure is a useful means of specifying the sizes of metal grains and is defined by N = (log n/log 2) + 1, where n is the number of grains per square inch at a magnification of 100×. Note that fine-grain metals have large ASTM numbers and coarse-grain metals have small ASTM numbers.

and continuous-cooling transformation (CT) diagrams. IT or TTT diagrams apply to transformations produced by holding the steel at a fixed temperature below the A_1 temperature and above the M_s temperature, whereas CT diagrams relate similar information for steel that is cooled continuously. Such plots are typically determined using very small specimens in which thermal equilibrium is attained rapidly throughout and temperature gradients are avoided.

Figure 2.9(a) shows a TTT diagram for a plain carbon steel containing 0.8% C. Because this steel is of the eutectoid composition, none of the austenite decomposes until the temperature of the steel has dropped below the eutectoid temperature (725 °C, or 1335 °F, in this steel). Notice that above the M_s temperature the TTT diagram assumes the form of two "C" curves. The left-hand C curve denotes the beginning of the transformation reaction, and the right-hand curve denotes its completion. The transformation is most rapid at an intermediate temperature between A_1 and M_s because pearlite and bainite form, for the most part, by two processes known as nucleation and growth. As its name implies, nucleation comprises formation of nuclei of the ferrite and carbide phases from austenite. Similarly, growth implies enlargement of the new nucleated phases. It turns out that nucleation is most rapid at temperatures considerably below the A_1 temperature at which the difference in internal energy between austenite on the one hand and ferrite and carbide on the other is great. In contrast, growth of the nucleated phase increases at higher temperatures at which the diffusion processes required for growth of new phases are enhanced. Because of the difference between nucleation and growth rates, it is, therefore, not surprising that the highest rate of formation of transformation products occurs at some intermediate temperature.

We have mentioned previously that, at higher transformation temperatures, pearlite is formed. Below the nose of the C curves, the harder ferrite-carbide microstructure known as bainite is produced by holding for sufficiently long times. Nevertheless, if the eutectoid steel is cooled very rapidly, avoiding the passage through the nose of the TTT diagram, or if it is cooled prior to complete transformation to pearlite or bainite, the austenite in the steel will transform to martensite when the M_s temperature is reached. Note that no holding time below M_s is required for such a reaction. This is because martensite is formed by a diffusionless reaction involving shear of the fcc lattice into a distorted bct lattice. The percentage of completion of the martensite reaction is determined solely by the temperature to which the steel is cooled. TTT diagrams often show the temperatures at which 50% or 90% of the austenite will have transformed (the M_{50} and M_{90} temperatures) or the temperature below which no more martensite forms (the martensite-finish or M_f temperature). The M_s and M_f temperatures decrease with increasing carbon content (Fig. 2.9b). In fact, for most steels with more than 0.6% C, M_f is below room temperature. Therefore, for these alloys, a small percentage of residual, or "retained," austenite is found even at room temperature after hardening. Cryogenic treatments are sometimes employed to transform this retained austenite, because it can seriously affect subsequent tempering behavior.

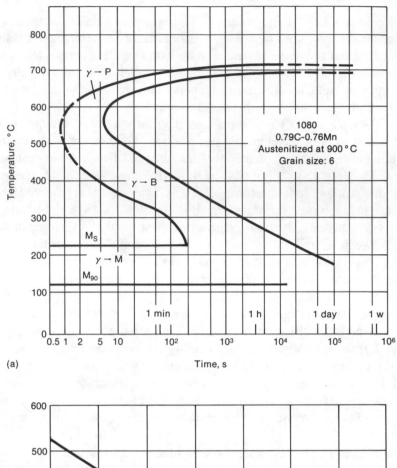

(a)

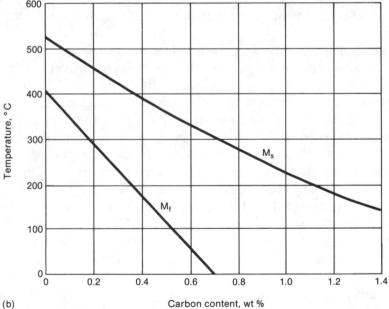

(b)

Fig. 2.9. (a) TTT diagram for a 0.79%C plain carbon steel
(P ≡ pearlite; B ≡ bainite; M ≡ martensite) (Ref 13). (b) M$_s$ and
M$_f$ temperatures as functions of carbon content for plain carbon
steels (Ref 14)

A transformation diagram somewhat different from that shown in Fig. 2.9(a) is obtained if the eutectoid steel is cooled continuously. The CT diagram for this material (Fig. 2.10) is similar to the TTT diagram except that the pearlite and bainite reactions are translated to longer times and lower temperatures. The CT diagrams for other steels bear a similar relationship to their TTT counterparts. The

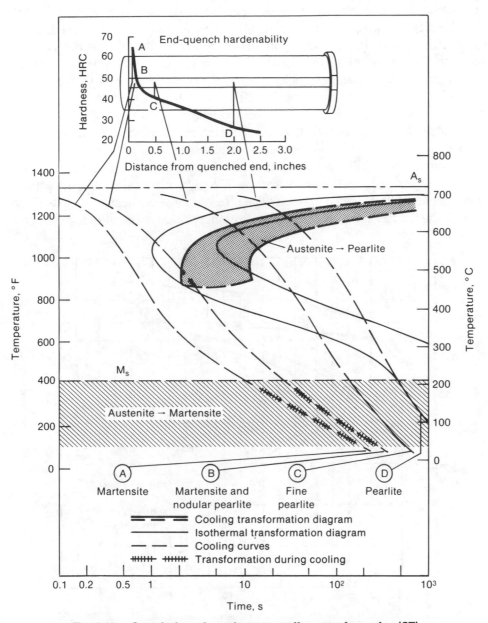

Fig. 2.10. Correlation of continuous-cooling transformation (CT) and isothermal transformation (IT) diagrams with end-quench hardenability test data for eutectoid carbon steel (Ref 15)

microstructures developed by continuous cooling are determined from these diagrams by tracing out the temperature-time history of the steel.

The TTT diagrams for other noneutectoid, plain carbon, and alloy steels are also similar to that given in Fig. 2.9(a) except that the decomposition of austenite above the A_1 temperature to produce ferrite (%C < 0.8) or carbide (%C > 0.8) is also included. Moreover, the temperature at which the nose is found, and the M_s and M_f temperatures, are different, depending on carbon content and other alloying elements present. An example for a typical line-pipe steel, AISI 1027, is shown in Fig. 2.11. In this illustration, the times for proeutectoid ferrite formation above the A_1 temperature (here called the A_s temperature) are given as well as those for pearlite and bainite, labeled only by the abbreviation F + C (ferrite plus carbide). Notice that the nose of the TTT curve begins at times too short to measure. Thus, the formation of pearlite and bainite starts very rapidly once the temperature falls to about 540 °C (1000 °F). Hence, it is practically impossible to form a totally martensitic microstructure in this steel.

To enhance the ability to avoid austenite decomposition before the M_s temperature is reached, and thus to achieve a totally martensitic microstructure, alloying elements are added to steels. The effect of these elements is to "push" the nose of the TTT curve to the right (i.e., to longer times). The alloying elements

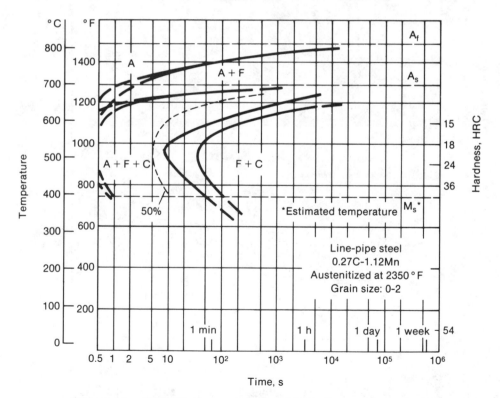

Fig. 2.11. TTT diagram for a 0.27%C line-pipe steel (A ≡ austenite; F ≡ ferrite; C ≡ carbide) (Ref 16)

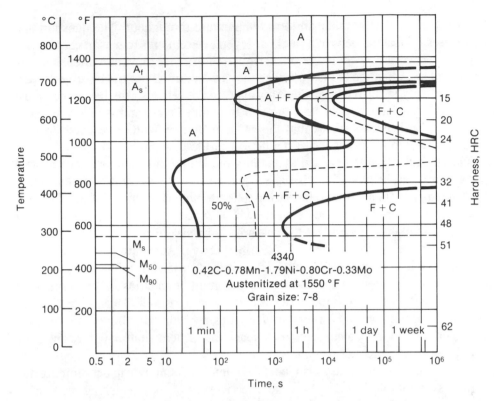

Fig. 2.12. TTT diagram for AISI 4340 steel (A ≡ austenite; F ≡ ferrite; C ≡ carbide) (Ref 17)

usually selected are ones which dissolve in the austenite in a substitutional form. Unlike carbon, which dissolves in the interstices of the fcc lattice of iron atoms because of its small atomic size, the usually larger substitutional elements take the place of some of the iron atoms themselves, leading to the term "substitutional." When austenite is cooled below the critical temperature, interstitial carbon can diffuse rather quickly, allowing for its rapid partitioning between ferrite and carbide phases. The partitioning of substitutional elements, on the other hand, is much more sluggish, because the diffusion processes in this case are much slower. The TTT diagram for the common alloy steel 4340, which contains substitutional additions of nickel, chromium, and molybdenum, is shown in Fig. 2.12. It can be seen that the pearlite reaction (the upper nose) is substantially delayed. The bainite reaction, however, is not retarded as much because it involves aspects both of diffusion, mentioned previously, and of some of the diffusionless, shear-transformation traits of martensite formation.

HARDENABILITY

Up to now, our discussions have centered on the transformation of austenite to pearlite, bainite, and martensite in extremely small steel specimens, which may be cooled rapidly to a rather uniform temperature throughout. The question that now arises is what happens to a steel bar several centimetres to tens of centimetres in diameter when it is cooled after austenitizing in order to form martensite. It is apparent that, when the bar is quenched into a water or oil bath, the surface layers are most likely to form martensite. However, in the center of the bar, which must be cooled by conduction of heat through the outer layers and which thus cools more slowly, it is less likely that the M_s temperature will be reached prior to transformation of some of the austenite to pearlite or bainite.

The characteristic that is used to quantify the ability to form martensite in steel parts of various section sizes is called "hardenability." As might be expected, hardenability bears a strong relationship to the temporal positions of the TTT and CT diagrams for the particular steel, which in turn depend on the chemical composition of the steel *and* on the austenite grain size. The influence of alloying elements on the diagrams has been described above. Austenite grain size plays a role through its influence on the nucleation of the ferrite and carbide phases. Since these phases are often nucleated on austenite grain boundaries, large grains with fewer nucleation sites tend to delay austenite decomposition. Thus, both alloying and large austenite grain size tend to delay this reaction and allow more time for the inner areas of thick steel sections to reach the M_s temperature before austenite is transformed to pearlite or bainite.

A third factor that must also be taken into account when determining the hardness achievable in practice is the effectiveness of the quenching medium in removing heat from steel. Three media in particular are commonly used: oil, water, and brine. Of the three, oil tends to provide the slowest cooling of steel parts, and brine the most rapid cooling. Furthermore, the effectiveness of each of these liquids increases if it is agitated. Sometimes it is desirable to use a quenching medium whose severity is between that of water and oil. For this purpose, the so-called polymer quenchants are put to use. These include polyvinyl alcohol (the most common), polyalkylene glycol ethers, polyvinylpyrrolidone, and polyacrylates.

There are two major methods for determining and interpreting the hardenability characteristics of steels. These are the end-quench test and the concept of ideal critical diameter. The end-quench test, often called the Jominy test after its originator, consists of cooling of the end of an austenitized steel bar using a well-controlled spray of water (Fig. 2.13). After cooling to room temperature, two diametrically opposed, axial strips or flats are ground onto the hardened bar. Hardness measurements using a standard test such as the Rockwell C indentation method (to obtain profiles of the so-called "HRC" hardness) or the diamond pyramid hardness test (to obtain "DPH" hardness measurements) are then made. The hardness distribution is plotted as a function of distance from the quenched end. As examples, data of this sort for a plain carbon steel (1045) and for an alloy steel of slightly lower carbon content (4340) are presented in Fig. 2.14. Here it can be seen that at the surface, where 100% martensite has been formed in both steels,

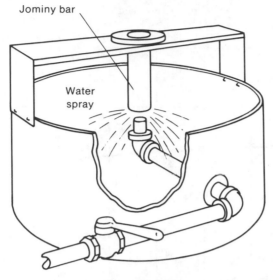

Schematic illustration of
Jominy test

Fig. 2.13.   The Jominy end-quench hardenability test (Ref 6, 18, 19)

the 1045 steel is slightly harder than the 4340 steel. This is because the hardness of martensite is determined primarily by carbon content, a fact which will be discussed further subsequently. Despite having the softer surface, however, the 4340 steel achieved a substantially higher hardness at depths well below the surface, and is therefore said to have higher hardenability. This can be ascribed to the presence of alloying additions of nickel, chromium, and molybdenum in this grade.

Another important feature of Jominy curves such as those shown in Fig. 2.14 is that they can be plotted as a function of hardness versus *cooling rate* as well

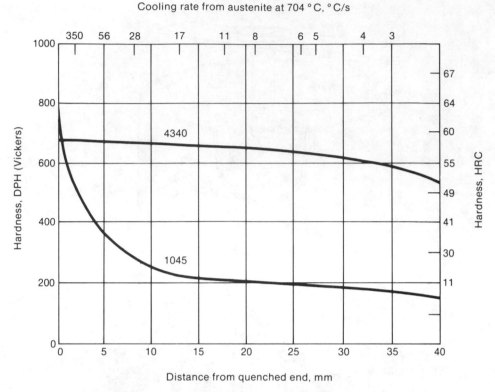

Cooling rate from austenite at 704 °C, °C/s

Fig. 2.14. Jominy end-quench hardenability curves for a 1045 steel (0.47C-0.75Mn-0.24Si) and a 4340 steel (0.39C-0.74Mn-0.31Si-1.73Ni-0.87Cr-0.24Mo) (Ref 20)

as of distance from the quenched end. Because of the standardized test geometry and the fact that most steels have similar thermal properties, the cooling rate through the transformation region bears a fixed relation to the distance from the quenched end. As a result, hardness data can be correlated with the micro-structures developed and predicted by CT curves. An example is given in Fig. 2.10 for type 1080 plain carbon steel. The cooling rate (through the trans-formation region) at a position only several millimetres below the surface is several hundred degrees Celsius per second, and the nose of the transformation is easily avoided. The microstructure formed is totally martensitic. At approximately 12 cm (4.72 in.) from the quenched end, however, the cooling rate is only about 17 °C/s (31 °F/s), and a considerably softer pearlitic microstructure is formed.

A somewhat more precise definition of hardenability is obtained through the concept of the ideal critical diameter, or D_i, developed originally by Grossmann and Bain. The ideal critical diameter is the diameter of a steel bar which when cooled at an infinitely fast rate will yield a microstructure of 50% martensite at the center. The 50% figure is chosen because it is much easier to establish such a microstructure metallographically than one containing 100% martensite. The ideal critical diameter is a material property which depends only on the chemical composition of the steel and the austenite grain size prior to quenching. The effects of these variables have been determined experimentally by a very large number of experiments in which each has been varied. Figure 2.15 gives the results of some of these experiments and may be used to calculate D_i values for actual steels. The calculation is begun by referring to the top part of the figure in which the critical diameter pertaining to the carbon content and austenite grain size of the steel under consideration is found. This number is then multiplied by a series of factors, one for each alloying element, whose magnitudes are based on the weight percentages of those elements in the steel. For instance, a commercial 1040 steel with an austenite grain size of ASTM 7, and a composition of 0.40% C, 0.83% Mn, and 0.31% Si, has an ideal critical diameter of (0.63) (1.6) (1.2) = 1.21 in. = 3.07 cm. By way of comparison, a 4140 steel with the identical austenite grain size and identical carbon, manganese, and silicon contents, but with alloying additions of 0.20% Ni, 1.00% Cr, and 0.19% Mo, would have a D_i of $(0.63)(1.6)(1.2)(1.2)(2.8)(1.2)$ = 4.9 in. = 12.4 cm, or a four-fold increase in hardenability.

The actual diameter which can be quenched to 50% martensite at the center in practice, or D_o, may be determined from another set of figures which enable the severity of the quenching medium to be taken into account. A perfect quenchant is said to have a quench severity coefficient (H value) of infinity. For this perfect quenchant, D_o is equal to D_i. In contrast, actual oil, water, and brine quenches have finite H values, as shown in Table 2.2, and in these real situations D_o is less than D_i. With these H values and the theoretical D_i value, the plots in Fig. 2.16 can be employed to obtain D_o values by merely reading up from the D_i abscissa to the appropriate H curve and then reading across to obtain D_o.

Table 2.3 presents the estimated maximum diameters at which cylinders made of a series of steels containing 0.40 to 0.50% carbon can be hardened to 50 HRC at the center (corresponding approximately to a structure of 50% martensite) by oil and water quenching. It is obvious that both the hardenability of the steel and the quenching medium used play significant roles in the hardening process.

Table 2.2. Quench severity coefficients (H values) for oil, water, and brine quenches

Degree of agitation	H value for: Oil	Water	Brine
None	0.25 to 0.30	0.9 to 1.0	2.0
Mild.	0.30 to 0.35	1.0 to 1.1	2.0 to 2.2
Moderate	0.35 to 0.40	1.2 to 1.3	. . .
Good	0.4 to 0.5	1.4 to 1.5	. . .
Strong	0.5 to 0.8	1.6 to 2.0	. . .
Violent.	0.8 to 1.1	4.0	5.0

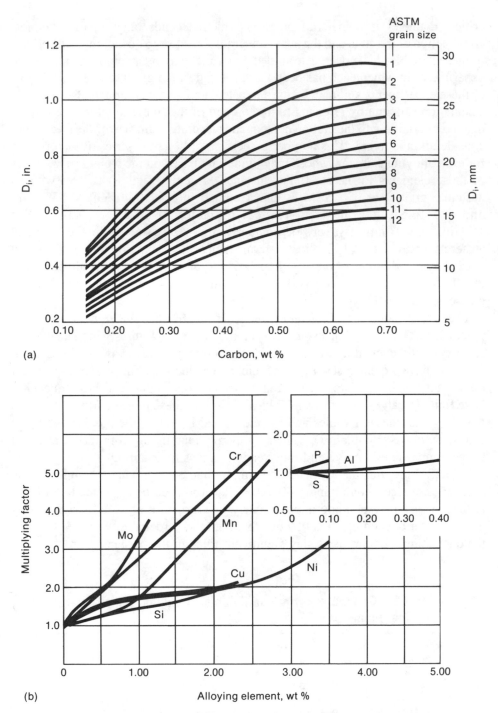

Fig. 2.15. (a) Ideal critical diameter, D_i, as a function of carbon content and austenite grain size and (b) multiplying factors for various alloying additions (Ref 21)

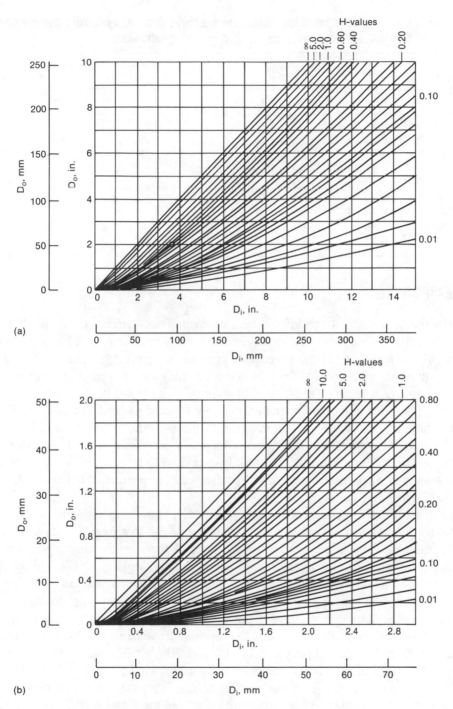

The lower diagram is an enlargement of the lower left-hand portion of the upper diagram.

Fig. 2.16.   Correlation between critical diameter, D_o, and ideal critical diameter, D_i, for quench media and degrees of agitation characterized by various H values (Ref 22)

Table 2.3.   Estimated maximum diameters at which steel cylinders can be hardened to 50 HRC at the center by oil and water quenching

AISI/ SAE No.	Composition, %						Maximum diameter			
							Oil quenching		Water quenching	
	C	Mn	Ni	Cr	Mo	V	cm	in.	cm	in.
1050	0.50	0.75	...	...	...	...	......	0.76	0.30	1.78	0.70
2340	0.40	0.75	3.50	...	...	...	......	4.57	1.80	6.35	2.50
3145	0.45	0.75	1.25	0.60	...	...	......	2.79	1.10	4.57	1.80
3240	0.40	0.45	1.75	1.00	...	...	......	5.84	2.30	6.86	2.70
3340	0.40	0.45	3.50	1.50	...	...	......	12.7+	5.00+	12.7+	5.00+
4140	0.40	0.75	...	0.95	0.20	...	......	4.57	1.80	6.60	2.60
4340	0.40	0.65	1.75	0.65	0.35	...	......	8.64	3.40	12.7	5.00
5140	0.40	0.75	...	0.95	...	...	......	2.29	0.90	3.56	1.40
6140	0.40	0.75	...	0.95	...	0.2	......	2.03	0.80	4.06	1.60

TEMPERING

In the as-quenched condition, the martensitic microstructure, although very hard, lacks the toughness often required for many applications. This is partially due to the fact that the microstructure contains a very large number of defects known as "dislocations," which are nonuniformities in the stacking arrangement of the atoms in the crystal lattice. This type of defect (and its density) is similar to what is found in the microstructures of heavily cold worked metals of similarly low ductility and toughness. In addition to the presence of the dislocation substructure, the presence of the large nonequilibrium concentration of carbon in solution impedes deformation and also leads to low toughness. In order to restore toughness to martensite, tempering is performed at temperatures between 100 and 700 °C (210 and 1290 °F).

During tempering, and depending on the exact tempering temperature, several microstructural changes occur, most of which give rise to a greater or lesser degree of softening. The major feature of all such changes is the decomposition of martensite into a fine microstructure of ferrite and carbide (Fig. 2.17). This occurs by a "precipitation" reaction in much the same way that excess sugar (or a "supersaturation" of sugar) may be made to precipitate from a water solution. The carbides that form are of three types: epsilon (ϵ), Fe_3C, and complex metal carbides. In plain carbon steels, epsilon carbides come from solution at tempering temperatures between approximately 100 and 200 °C (210 and 390 °F). These so-called transition carbides come from solution within the fine platelets that characterize martensite. Because they are so fine, they may sometimes even give rise to slight *increases* in hardness, particularly in the higher-carbon steels. This effect is shown for a variety of plain carbon steels in Fig. 2.18. At temperatures between 250 and 700 °C (480 and 1290 °F), the Fe_3C carbides come from solution and grow in size to produce softening the magnitude of which increases with

(a)

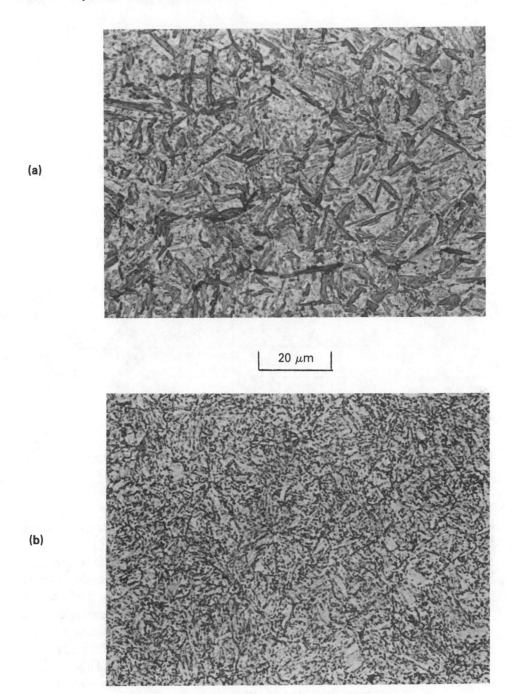

|— 20 μm —|

(b)

Fig. 2.17. Microstructures of (a) as-quenched martensite and (b) tempered martensite (tempered 10 min at 675 °C, or 1250 °F) in a 1045 steel

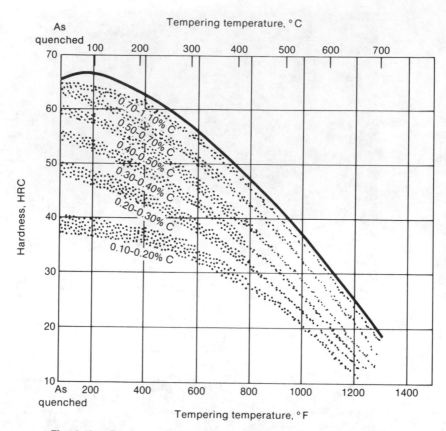

Fig. 2.18. Decrease in hardness with increasing tempering temperature (tempering time, 1 h) for carbon steels of various carbon contents (Ref 23)

temperature. The softening is quite marked for the plain carbon steels shown in Fig. 2.18, which were tempered for 1 h at temperature.

Much like the transformation reactions that characterize pearlite and bainite formation, the tempering process in alloy steels is considerably slower than in their plain carbon counterparts. As before, this is due to the retarding effects of alloying elements on diffusion processes. Hence, for a given tempering temperature and time, alloy steels quenched to the same hardness may be expected to have a *higher* tempered hardness. Another effect which tends to lead to higher tempered hardnesses in alloy steels is the phenomenon of "secondary hardening." This phenomenon occurs at around 500 to 600 °C (930 to 1110 °F) when alloy carbides are precipitated from solution. These carbides may lead to hardness peaks in plots of tempered hardness versus tempering temperature such as the one shown in Fig. 2.19. As might be expected, the amount by which softening is retarded and the level of the secondary hardening peak are greatly influenced by the amount of alloying addition, a fact illustrated in Fig. 2.20 by data for steels of various molybdenum contents.

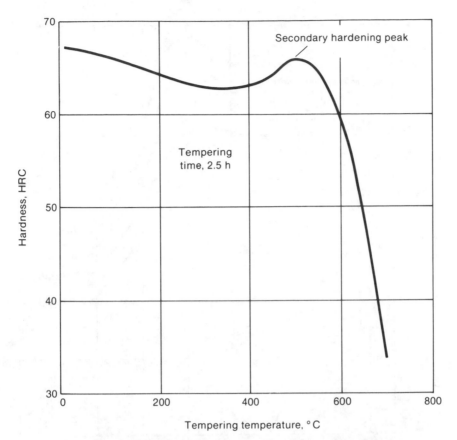

Fig. 2.19.   Tempering curve for a 0.7C-18W-4Cr-1V tool steel austenitized at 1280 °C (2335 °F), oil quenched, and tempered for 2.5 h (Ref 24)

Other solid-state reactions occur when retained austenite or transformation products from hardening treatments (pearlite and bainite) are tempered. During tempering, retained austenite transforms to pearlite or bainite depending on the tempering temperature. This pearlite and bainite, as well as those contained in the microstructure following the hardening treatment, are also softened by tempering treatments. This is a result of the general coarsening and spheroidization of the Fe_3C constituent in these microstructures.

Some steels may require two tempering treatments, or "double tempering." This is often the case when such a steel is austenitized and quenched to a temperature slightly above (e.g., 50 to 75 °C, or 90 to 135 °F, above) the M_f temperature in order to prevent quench cracking. If tempering is subsequently conducted at a relatively low temperature, the retained austenite may not transform. Upon cooling to room temperature, *untempered martensite* is formed from this retained austenite. A second tempering treatment is then preferred, usually at a temperature slightly below the first, to avoid the low fracture resistance associated with such a microstructure.

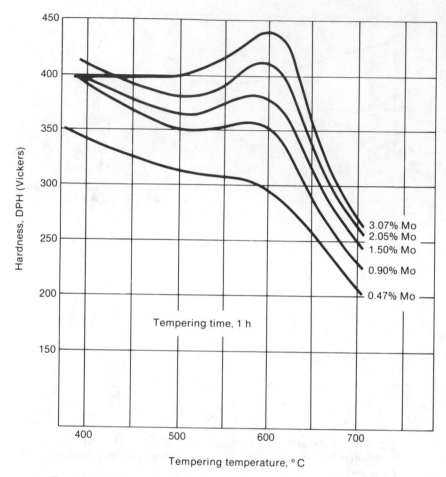

Fig. 2.20.   Retardation of softening and the occurrence of secondary hardening during tempering of steels containing various amounts of molybdenum (Ref 25)

TIME-TEMPERATURE RELATIONSHIPS FOR TEMPERING

Conventional tempering operations are conducted in gas-fired furnaces using times on the order of one to several hours. Often the desired final hardness is known, and curves such as those in Fig. 2.18 and 2.19 (for fixed tempering time) are consulted to determine the temperature required. Because tempering reactions occur at different rates at different temperatures (giving rise to the hardness variation in tempering curves), however, equivalent final tempering results may be obtained by a wide range of combinations of time and temperature. A given hardness achieved after long-time tempering at a low temperature can often be achieved just as easily by short-time tempering at a higher temperature. This point is illustrated by the data for 4340 steel in Fig. 2.21. Here, for example, it can be

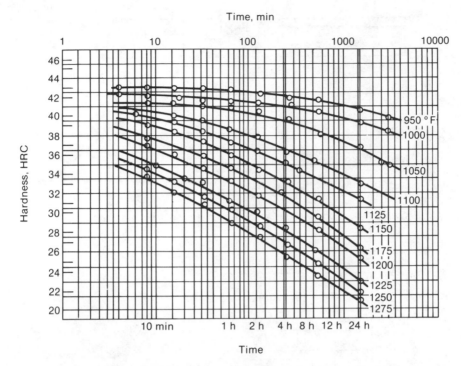

Fig. 2.21. Tempering behavior of a 4340 steel (0.355C-0.66Mn-0.024P-0.017S-0.28Si-0.70Cr-1.58Ni-0.37Mo) at various temperatures as a function of time (Ref 26)

seen that a hardness of 35 HRC is achieved by tempering for 10 min at 665 °C (1225 °F) as well as for 100 min at 620 °C (1150 °F).

Since tempering of martensite is such an important heat treatment operation, a great deal of work has gone into establishing the time-temperature relationships that will result in identical hardness values for a variety of steels. Most of these relations show that hardness is a function of the product of the absolute tempering temperature times the sum of a constant and the common logarithm of the tempering time, or:

$$\text{Hardness} = \text{Function of } T(C + \log t)$$

where T is the temperature in kelvins or degrees Rankine, t is the time in seconds, and C is a constant that depends on alloy composition. This relation was first proposed by Hollomon and Jaffe (Ref 27), who found that the value of the constant C was generally between 10 and 15 for the steels in their study. Figure 2.22 shows the very good fit that the above equation yields for a plain carbon steel and an alloy steel. Other examples of the application of this relation are illustrated in Fig. 2.23 and 2.24, for which the fit can be seen to be equally as good with the proper choice of the constant C.

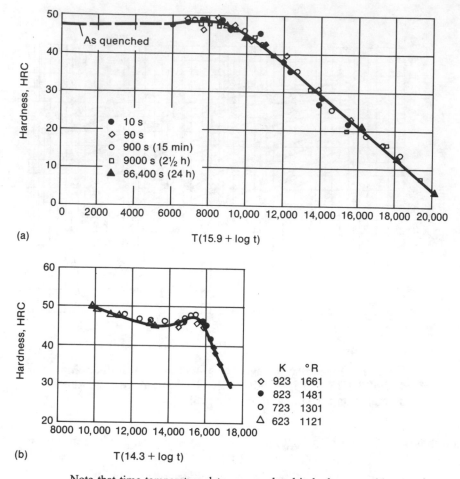

$$T(15.9 + \log t)$$

$$T(14.3 + \log t)$$

Note that time-temperature data are correlated in both cases with a parameter of the form $T(C + \log_{10} t)$, where T is absolute temperature in K, C is a constant, and t is time in seconds.

Fig. 2.22.   Tempering curves for (a) a 0.31%C plain carbon steel and (b) a 0.35C-2Mo alloy steel which exhibits secondary hardening (Ref 27)

Hollomon and Jaffe were also successful in demonstrating that the kinetics, or the time-temperature relations, for tempering of pearlite and bainite may be described by an identical relation. The data in Fig. 2.25, also taken from their work, suggest that the value of the constant C is the same irrespective of microstructure.

The data just presented offer strong support for a unique dependence of tempered hardness on the Hollomon-Jaffe time-temperature parameter. This is probably true for most steels except those in which secondary hardening occurs. In these cases, the C constants may vary for the individual processes of Fe_3C precipitation and alloy carbide precipitation, which in turn vary with temperature in different ways. For this reason, a unique value of C may not be obtainable. However, there

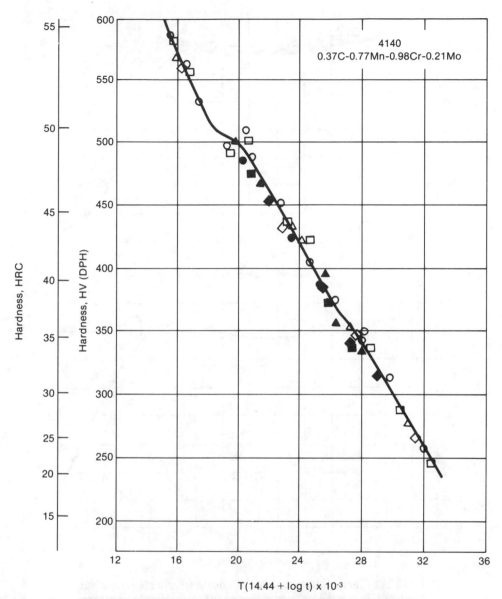

Fig. 2.23. Tempering curves for 4140 steel based on the parameter $T(14.44 + \log_{10} t)$, where T is absolute temperature in °R and t is tempering time in seconds (Ref 28)

are cases in which these considerations are inapplicable, examples of which are given in Fig. 2.22(b) and 2.24, both of which show secondary hardening peaks.

PROPERTIES

In previous sections, the heat treatment of steel has been discussed primarily in terms of the phase changes that occur with little attention paid to the properties

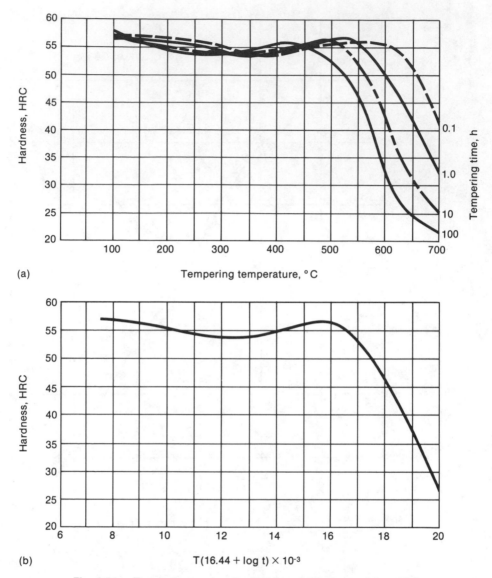

Fig. 2.24.   Tempering curves for H13 hot work die steel in which hardness is plotted as a function of (a) tempering temperature (for various tempering times) and (b) the time-temperature tempering parameter, $T(16.44 + \log_{10} t)$, where T is absolute temperature in K and t is tempering time in seconds (Ref 5)

(except hardness) that are developed. In this section, the important characteristics that are controlled by heat treatment will be discussed. These characteristics are hardness, tensile properties, toughness, and fatigue resistance.

Hardness

The first property, hardness, has already been touched on briefly. This property is a measure of wear resistance and the ability of a material to support loads in

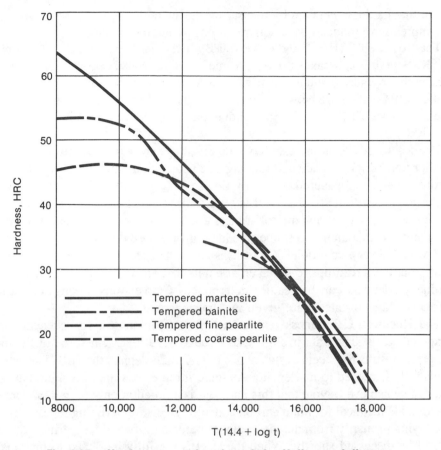

Fig. 2.25. Hardness as a function of the Hollomon-Jaffe tempering parameter (T in K, t in seconds) for a 0.94%C plain carbon steel with starting microstructures of martensite, bainite, fine pearlite, and coarse pearlite (Ref 27)

service without suffering permanent shape change, or plastic deformation. Typically, hardness is measured in a standardized test in which an indenter is pressed into the steel using a prescribed load. The depth of penetration or the width or area of the hardness impression is used as the measure of hardness.

For steels, the most common hardness tests are those referred to by the names Brinell, Diamond Pyramid, Rockwell C, Rockwell B, and Knoop. In the Brinell test, a 10-mm (0.39-in.) diameter hardened steel or tungsten carbide ball is pressed into the test specimen under a load of 3000 kg (6600 lb). This leaves a fairly large impression, which is desirable when the hardness of a steel with a relatively coarse microstructure is to be established. From the surface area of the impression, the Brinell hardness number (HB) is determined; typical values are in the range of 100 to 750. Higher numbers indicate high hardness, and lower numbers low hardness. Also, the Brinell hardness number divided by two gives a good estimate of ultimate tensile strength (in thousands of pounds per square inch, or ksi). The Diamond Pyramid, or Vickers, hardness (DPH or HV) test is similar to the Brinell test in that hardness is taken by measuring the average length of the diagonals of the impression produced by a square-based, pyramidal indenter.

The Rockwell C (HRC) and Rockwell B (HRB) hardness tests are also similar. The Rockwell C test makes use of a "Brale" indenter consisting of a 120° diamond cone with a rounded point and is used to make measurements on the harder steels. Typical HRC values are between 20 and 70. The Rockwell B test is used for softer steels and uses a 1.59-mm (1/16-in.) diameter steel ball. HRB values range from 0 to 100.

The Rockwell hardness tests have an advantage over the Brinell test in that a smaller impression is made and readings are taken automatically. However, when hardness must be measured on a very fine scale (e.g., when the hardness of one of the microconstituents of a steel is required), a so-called "microhardness" test is used. The most common microhardness test is the Knoop test. This test employs an indenter with a diamond shape, the lengths of whose diagonals are in the ratio of 7 to 1, and very small loads (as little as 25 g). The shape of the Knoop indenter allows hardness readings to be taken at points that are very close together so that hardness gradients can be readily determined. Conversions among the various hardness scales for steels are given in Appendix A.

The Rockwell C hardness test is the most common test for quenched-and-tempered steels. The data in Fig. 2.26 through 2.29 represent some of the more important results for steels of this class. Figure 2.26 depicts the HRC hardness of steels fully hardened to martensite. As such, it represents the maximum hardness that can be achieved. Note that this hardness is a function only of carbon content. The trend is identical for plain carbon and alloy steels, underlining the importance of alloying primarily from the viewpoint of hardenability and tempering behavior, as will be discussed shortly. When the steel is not fully hardened to martensite, the hardness drops, and the magnitude of this lower hardness is also a function of carbon content (Fig. 2.27). One of the important features about Fig. 2.26 and

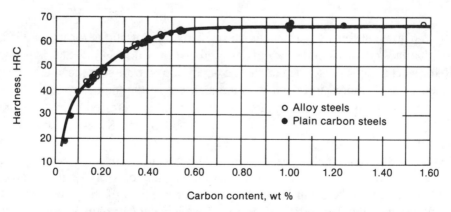

Fig. 2.26. Hardness of as-quenched martensite as a function of carbon content (Ref 29, 30)

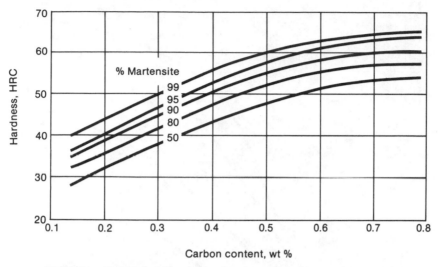

Fig. 2.27. **Relationship among hardness, carbon content, and amount of martensite (Ref 31)**

2.27 is the fact that the hardness reaches a plateau at a carbon content of about 0.6 to 0.8%.

Following hardening, the hardness of a steel may change a great deal depending on tempering temperature. Figure 2.28 shows this effect for a variety of plain carbon steels in terms of both Rockwell C (HRC) and Vickers (HV) numbers. Here, the change of hardness with tempering temperature (for a tempering time of 1 h) is greatest for the higher-carbon alloys. Such a trend might have been expected since the tempering process consists of the precipitation and growth of carbides. Steels with higher carbon contents thus lose more strength (in terms of HRC points) because of carbide coarsening.

The addition of alloying elements, though not helpful from the standpoint of martensite hardness, does tend to enable higher tempered hardnesses to be achieved than are possible in plain carbon steels. This effect is illustrated in Fig. 2.29, where the increase in hardness as a function of alloy content is shown for tempering treatments of 1 h at 540 °C (1000 °F). The tempered hardness is obtained by adding the base hardness of the corresponding plain carbon steel and hardness increment factors for each alloying element. The magnitudes of these increments are determined from the figure at the appropriate level of alloying. For tempering at 540 °C (1000 °F), strong carbide formers (V, Mo, Cr) give rise to substantial hardness increments; the effects of these elements are much less at lower tempering temperatures, however, since Fe_3C precipitation predominates in this regime. Nickel has little effect on tempering because it remains in solid solution. In contrast, manganese has a large influence at 540 °C (1000 °F), presumably because of its ability to retard coarsening of cementite. The detailed effects of alloying on tempering behavior for temperatures between 205 and

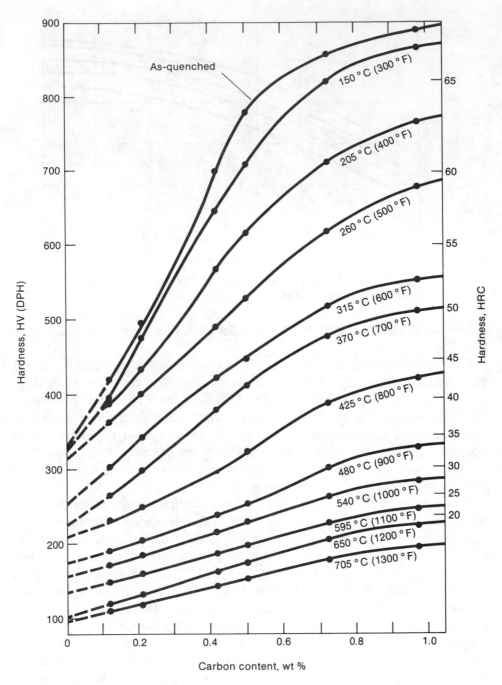

Fig. 2.28.   Hardness as a function of the carbon content of martensite in iron-carbon alloys tempered at various temperatures (Ref 32)

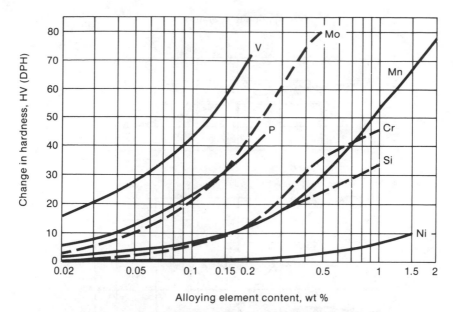

Fig. 2.29. Increases in tempered hardness (relative to iron-carbon alloys) resulting from addition of various alloying elements for alloys tempered at 540 °C (1000 °F) for 1 h (Ref 32)

705 °C (400 and 1300 °F) are described in a paper by Grange, Hribal, and Porter (Ref 32).

Tensile Properties

Tensile properties also give an indication of the ability of a material to withstand loads in service and are derived from tests in which bars of the material are pulled uniaxially to failure. From these experiments, measurements of yield strength, ultimate tensile strength, elongation, and reduction in area are obtained. Yield strength is the stress (load divided by cross-sectional area over which the load acts) under which a metal first suffers plastic deformation. Similarly, ultimate tensile strength (or UTS), often simply referred to as "tensile strength," is the stress (based on the initial cross-sectional area of the tensile specimen) at which the load during testing achieves a maximum. The units used for both yield strength and tensile strength are pounds per square inch (psi; 1000 psi = 1 ksi) and newtons per square millimetre (N/mm^2), to which the term "pascal" (Pa) is applied. High values of these strength parameters, particularly yield strength, are useful for service applications.

Two ductility properties are also commonly obtained in tensile tests: elongation and reduction in area at fracture. The first of these is the percentage increase in length of the gage section of the tensile specimen at fracture. High values of elongation imply a resistance to brittle fracture, although for high-strength grades of steel the reduction-in-area (RA) parameter is probably a better measure. As its

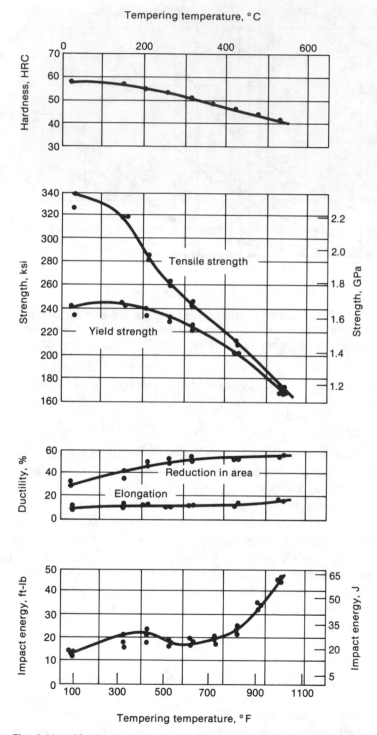

Fig. 2.30.   Mechanical properties of a 4340 steel oil quenched to produce a martensitic structure and tempered for 1 h at various temperatures (Ref 33)

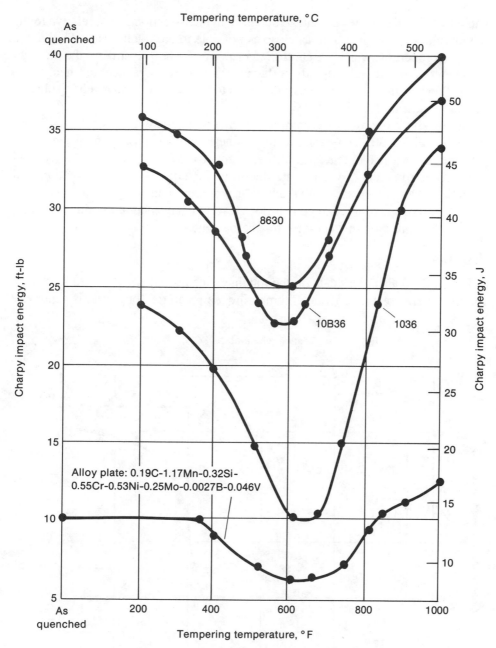

The alloy plate was tested at −30 °C (−20 °F); the other three steels were tested at room temperature. Note the decrease in toughness for tempering temperatures from 230 to 370 °C (450 to 700 °F).

Fig. 2.31. Charpy impact strength versus tempering temperature for various quenched-and-tempered steels (Ref 34)

name implies, RA is the percentage decrease in cross-sectional area relative to the initial value, and is measured as close as possible to the fracture site. Unlike the strength properties, which are controlled primarily by heat treatment, the ductility parameters are also substantially influenced by melting practice and heat treatment. Poor melting practice, which may lead to "dirty" steel, can result in substantial reductions in ductility.

The tensile properties of a common quenched-and-tempered steel, 4340, are shown in Fig. 2.30. The decrease in hardness with increasing tempering temperature, shown also in this figure, is accompanied by a general decrease in yield and tensile strengths. By contrast, ductility, in terms of both elongation and reduction in area, increases with the drop in strength as the tempering temperature is increased. The observed trend is followed for all quenched-and-tempered steels and poses a question of trade-offs for the designer.

Toughness

Figure 2.30 also gives data on the toughness or impact energy of 4340 steel at room temperature as a function of tempering temperature. Toughness is a measure

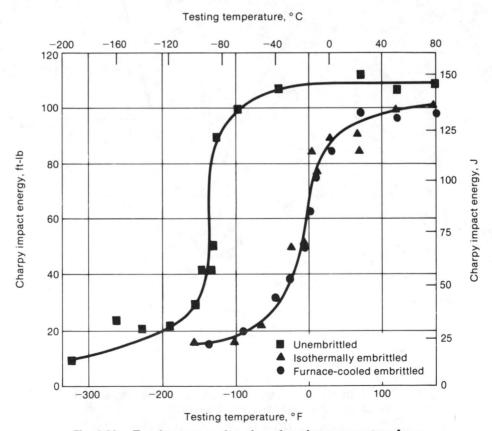

Fig. 2.32.   Toughness as a function of testing temperature for a steel subject to temper brittleness in both embrittled and unembrittled conditions (Ref 35)

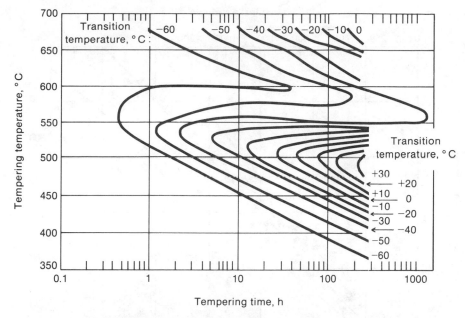

The steel (0.4C-0.8Mn-0.6Cr-1.25Ni) was water quenched from 900 °C (1650 °F) and double tempered—first for 1 h at 675 °C (1250 °F) and then for the times and temperatures indicated.

Fig. 2.33. Dependence of the ductile-to-brittle transition temperature on tempering time and temperature for quenched-and-double-tempered 3140 steel (Ref 36)

of the ability of a material to resist brittle fracture under load in the presence of flaws (gouges, welding defects, and so forth). Probably the simplest means of estimating its magnitude for various steels is the Charpy test. Named for the French engineer who designed it, the Charpy test consists of fracturing a notched test bar using a swinging pendulum. The "impact" energy required to accomplish this, readily determined from the height to which the pendulum swings following fracture, is used as the measure of toughness.

Curves of impact energy for standard 4340 Charpy specimens reveal an interesting trend. This energy increases at first with tempering temperature, passes through a minimum, and then increases again. The minimum is due to 260 °C (500 °F) embrittlement, a name given to this effect because of the tempering temperature which produces it. At this tempering temperature, "tramp" elements such as lead, antimony, and tin, which are present in small yet unavoidable quantities in conventionally refined steels, segregate to the grain boundaries during the long-time tempering treatments given to quench-hardened martensitic microstructures. The precipitation of Fe_3C in platelet form at these tempering temperatures is also partly responsible for the embrittlement. Tensile ductility is not affected by this phenomenon; there are no minima in the plots of elongation and reduction in area versus tempering temperature. Only the toughness is affected, and this effect occurs only when toughness is measured at ambient or

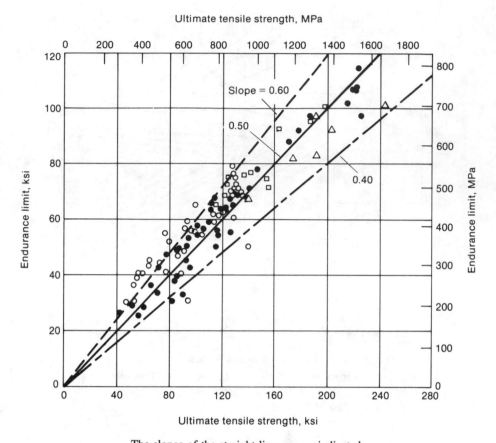

The slopes of the straight lines are as indicated.

Fig. 2.34.   Relationship between fatigue endurance limit and tensile strength for several steels (Ref 37)

cryogenic temperatures. Toughness data for other quenched-and-tempered steels, showing a similar effect of tempering temperature, are given in Fig. 2.31.

A related embrittlement problem often called "temper brittleness" results from tempering martensite in, or slow cooling it through, the temperature range of 400 to 550 °C (750 to 1020 °F). As with 260 °C (500 °F) embrittlement, temper brittleness is manifested by room- or low-temperature brittle fracture, which is detectable by Charpy testing but not by tensile testing. The best way to determine the losses in properties that accompany such tempering treatments is by performing Charpy tests at various temperatures. In an embrittled steel, the ductile-to-brittle transition temperature (DBTT), or the temperature at which the impact energy drops sharply, will be much higher than in the unembrittled condition. This is easily detected in plots of impact energy versus test temperature, such as those in Fig. 2.32. Here, the embrittling effect raised the DBTT (or, simply, the "transition temperature") from −95 to −20 °C (−140 to −5 °F). This may not seem important until it is realized that many steels in the unembrittled condition have transition temperatures near room temperature.

The temperature range in which temper brittleness is induced (400 to 550 °C, or 750 to 1020 °F) appears to be quite large. Fortunately, the effects are great only at temperatures near the midpoint of this range and when long-time tempering is used. This fact is illustrated for 3140 steel in Fig. 2.33, which shows that tempering times in excess of 1 h are normally required for large increases in the transition temperature.

Fatigue Resistance

The last property to be discussed in this chapter is fatigue resistance. Fatigue is probably the most common failure mechanism of steels during service operation. It occurs by slow growth of a crack which is usually initiated at a weak point on the surface of the structure or at a region where the loads are higher than elsewhere (a "stress concentration"). Fatigue cracks can initiate and grow even when the applied loads result in stresses below the yield strength. However, many metals are characterized by a threshold, or "endurance limit," below which fatigue

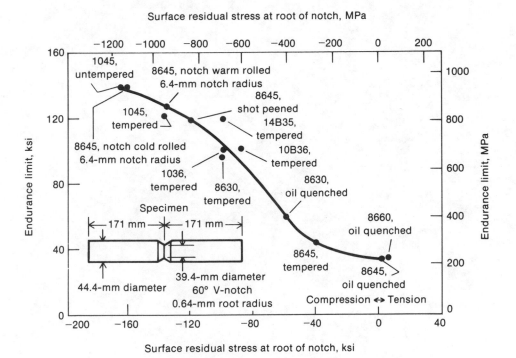

All steels were water quenched except where otherwise indicated.

Fig. 2.35. Relationship between bending fatigue endurance limit and surface residual stresses for various carbon and alloy steels (Ref 34)

cracking can be avoided. For quenched-and-tempered steels, this limit is about one-half the yield strength (Fig. 2.34). Another important point about fatigue is that the resistance to it can be increased by inducing *compressive* residual stresses on parts subject to failure by this mechanism. This behavior is a result of the fact that fatigue is initiated under the action of tensile stresses. Thus, superimposing a compressive residual stress pattern may partially or totally eliminate the surface tensile stress component and allow higher *applied* tensile loads to be supported without reaching the endurance limit. This effect is illustrated by the results shown in Fig. 2.35.

Chapter 3

Fundamental Principles of Induction Heating

There are many texts concerning the principles of induction heating. These books are structured primarily for electrical engineers and discuss in detail the derivations of various pertinent electrical relationships and the design of induction equipment. On the other hand, there are short articles in the technical literature which briefly explain the general principles involved but do not furnish enough information for system design or calculation of important heating parameters. The treatment in this and the following chapter (on equipment and coil design) is meant to offer a compromise between the two extremes. It is intended to explain the general theory of induction heating and equipment design in order that engineers can design and construct routine induction heating systems, be they for steel heat treating or otherwise. For more in-depth discussions, the reader is referred to one of the full-length induction heating textbooks (Ref 38 to 40). A brief review of important electrical engineering principles is given in Appendix C.

BASIS OF INDUCTION HEATING

Induction heating provides a means for precise heating of electrically conducting objects. In some cases, it is the only practical method of supplying heat to the work material. It is clean, fast, and repeatable, and lends itself to automatic cycling. No contact is required between the work load and the heat source, and heat may be restricted to localized areas or surface zones.

The basis for induction heating lies in the ability to *induce* electric currents in electrical conductors. Two simple methods of inducing such currents are illustrated in Fig. 3.1. In Fig. 3.1(a), the terminals of a conductor are connected across a current-measuring device, or galvanometer. Normally, no current would be detected because there is no source of electromotive force (such as a battery) in the closed path of this circuit. If, however, a permanent magnet were moved through the wire loop, a current would be detected. If the magnet were moved away from the conductor, the galvanometer would deflect again. Further experi-

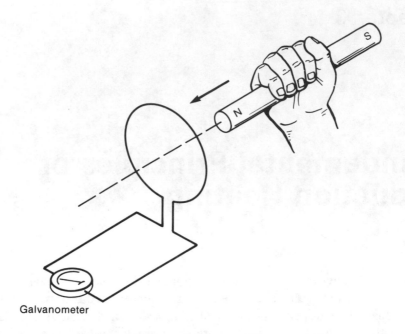

Galvanometer

(a)

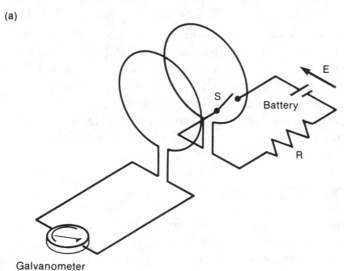

Galvanometer

(b)

In (a), electric current is induced while the magnet is moving with
respect to the coil. In (b), electric current is induced momentarily
when the switch S is opened or closed.

**Fig. 3.1.   Two simple experiments illustrating electromagnetic
induction (Ref 41)**

mentation would show that currents are produced only when there is relative motion between the conductor and the magnet. These currents are known as induced (or *eddy*) currents, and the principle illustrated here forms the basis for design of steam-powered electric generators.

A second means of inducing eddy currents is shown in Fig. 3.1(b). In this drawing, the conductor/galvanometer arrangement shown in Fig. 3.1(a) is placed beside an identical conductor which is connected in series with a resistor (of resistance R), a battery (of electromotive force E), and a switch. When the switch is closed, a current, I, whose value is given by Ohm's Law (I = E/R) is generated in the right-hand conductor. This current is called a direct current (dc) because its direction is fixed; it flows counterclockwise from the "high-potential" side of the battery (the longer "bar" in the battery symbol) around the circuit to the "low-potential" side of the battery. While this direct current is flowing, no current is detected in the adjacent conductor. However, when the switch is closed or opened, a momentary or transient current will be induced and will be detected by the galvanometer.

The apparently unrelated means of inducing currents in electrical conductors described above can be rationalized by realizing that a magnetic field is associated with electric currents. This magnetic field lies in a plane perpendicular to the conductor (Fig. 3.2). In the example in Fig. 3.1(a), a current is induced when the conductor cuts lines of the magnetic field, or flux. Similarly, in Fig. 3.1(b), the opening and closing of the switch causes the associated magnetic field around the right-hand conductor to change. In so doing, the left-hand conductor cuts through lines of flux during these instants in time. Thus, in both examples, currents are induced when lines of magnetic flux are cut by the conductor.

In practice, it is not feasible to induce currents in conductors by the method illustrated schematically in Fig. 3.1(b). Instead, the right-hand element is usually energized by an alternating current (ac). In this case, the current direction and magnitude vary with time, typically in a sinusoidal fashion. These current variations lead to continuous magnetic field variations which result in induced currents in the adjacent conductor, known as the "workpiece" or "load."

A simple prototype induction heating setup is shown in Fig. 3.3. Here, the inductor is a helical, or solenoidal, coil (generally known as an "induction coil") carrying an ac current which surrounds a tubular workpiece. The magnetic fields between the turns of the coil vanish because adjacent turns set up magnetic fields of opposite sign in these regions. Thus, the magnetic field lines surround the coil as shown, and, as their signs and magnitudes vary, eddy currents are induced as also shown in the figure; generally, the path of the induced current in the workpiece is parallel to that of the currents in the inductor.

The electric currents generated, as described above, lead to heating of the workpiece (and of the inductor). Associated with the current is a voltage drop, V, which, for a pure resistance, is given by Ohm's Law (V = IR). When a drop in potential occurs, electrical energy is converted into thermal energy, or heat. This

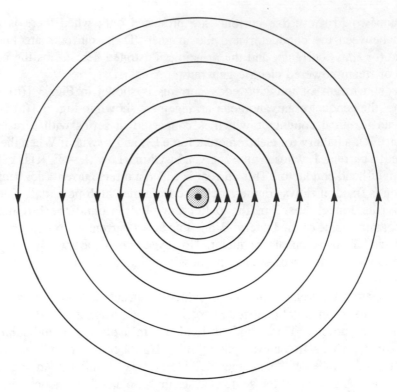

The current is emerging from the page. The relationship between
the directions of the magnetic field and the current is expressed by
the "right-hand" rule (thumb pointing in direction of current, fin-
gers giving direction of magnetic field).

**Fig. 3.2.   Magnetic field (lines with arrows) around an electrical
conductor (cross-hatched circle at center) carrying a current
(Ref 41)**

conversion of energy is analogous to the conversion of potential energy into
kinetic energy in mechanical systems, such as occurs when an object is dropped
under the force of gravity from a given height. In the electrical case, the drop in
voltage, or potential, results in heating at a rate given by $VI = I^2R$. Note that this
is a heating *rate,* or a measure of power—i.e., its units are energy/time.

If the workpiece is magnetic (e.g., steels at low temperatures), additional heat
is produced by hysteresis losses; this heating usually is small but, in some cases
involving strong magnetic fields, can become significant. Induction heating can
be contrasted to dielectric heating, in which nonconductors are heated by alternating electric currents.

In designing a total system for an induction heating application, the three major
steps involved are generally the following: (1) selecting a frequency and determining power requirements; (2) choosing a power source; and (3) designing an
induction coil. The first of these steps is described below; the others are discussed
in Chapter 4.

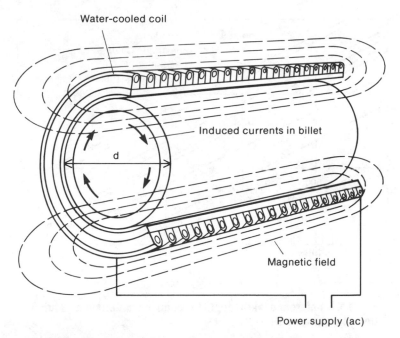

Fig. 3.3. Basic principle of induction heating illustrated for a tubular conductor heated by a solenoid coil (Ref 40)

FREQUENCY SELECTION

Skin Effect and Reference Depth

Induction heating is efficient and practical if certain basic relationships between the frequency of the magnetic field (and thus the frequency of the alternating current in the inductor) and the properties of the workpiece are satisfied. Although these relationships are not extremely critical, they must be satisfied to the extent that a suitable degree of skin effect is produced in the workpiece. Skin effect is the phenomenon by which the eddy currents flowing in a cylindrical workpiece tend to be most intense at the surface, while currents at the center are nearly zero. Similarly, for a flat workpiece, if energy is impinging from one side only, the eddy currents are greater at that surface. As a consequence of these types of current distributions, there exists a greater rate of heating near the surface. Skin effect is present in every successful induction heating application.

The mathematics needed to explain the skin effect are beyond the scope of this discussion, but it should be mentioned that the field intensity in a conductor that is impinged upon by an alternating electromagnetic field can be described by a differential equation that has solutions in the form of Bessel functions. These solutions demonstrate that the induced current decreases exponentially from the surface into the workpiece, or electrical load. For a thorough development of the mathematics of induction heating, the reader should consult books such as those by Brown, Hoyler, and Bierwirth (Ref 38) and Davies and Simpson (Ref 40).

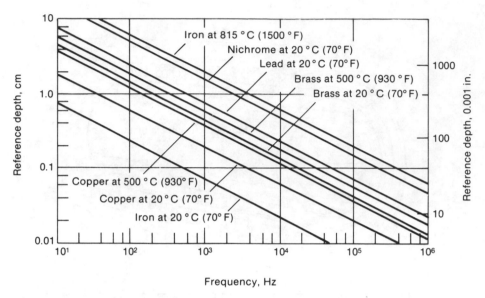

Fig. 3.4. Reference (skin) depth for common materials as a function of frequency (Ref 42)

One of the most important benefits obtained from the solution of the differential equation describing induction heating is that it allows us to define an "effective" depth of the current-carrying layers. This depth, which is known as the "reference depth" or "skin depth," d, depends on the frequency of the ac field and on the electrical resistivity and relative magnetic permeability of the workpiece, and is very useful in gaging the suitability of various materials for induction heating. The definition of d is:

$$d = 3160\sqrt{\rho/\mu f} \qquad \text{(English units)}$$

or

$$d = 5000\sqrt{\rho/\mu f} \qquad \text{(metric units)}$$

where d is the reference depth, in inches or centimetres; ρ is the resistivity of the workpiece, in ohm-inches or ohm-centimetres; μ is the relative magnetic permeability of the workpiece (dimensionless); and f is the frequency of the alternating magnetic field of the work coil, in hertz (cycles per second). The reference depth is the distance from the surface of a given material at which the induced field strength and current are reduced to $1/e$, or 37% of their surface values. The power density at this point is $1/e^2$, or 14% of its value at the surface.*

Figure 3.4 presents plots of reference depth versus frequency for various common metals. It varies with temperature, for a fixed frequency, because the re-

---

*e ≡ the base of the natural logarithm = 2.718.

sistivity of conductors varies with temperature. Furthermore, for magnetic steels, the permeability varies with temperature, decreasing to a value of one (relative to free space) at the so-called Curie temperature. Also, as the power density is increased, steels become magnetically saturated, leading to decreased permeability and hence increased reference depth (Fig. 3.5). Because of these effects, the

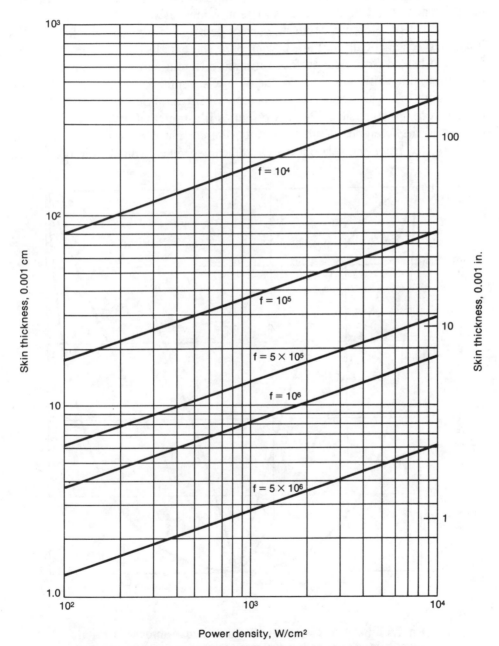

Fig. 3.5. Reference (skin) depth for magnetic steel as a function of power density and frequency (Ref 38)

reference depth in nonmagnetic materials may vary by a factor of two or three over a wide heating range, whereas for magnetic steels it can vary by a factor of 20.

An equivalent way of looking at reference depth is through the use of the equivalent sleeve concept. In a round bar, if all the current were concentrated uniformly in a sleeve at the surface rather than exponentially, and the heating were the same as in the actual case, the sleeve would be one reference depth thick, and its resistance would be the same as the equivalent resistance in the real case.

The effect of reference depth on current-density distribution in a flat sheet is illustrated schematically in Fig. 3.6 and 3.7. The data in Fig. 3.6 are for a sheet induction heated from both surfaces at three different current frequencies which give rise to three different values of the ratio of sheet thickness to reference depth, a/d. Figure 3.7 is also for induction heating of sheet, but in this case the sheet is heated from only one side. These two figures illustrate that if the material is thin relative to the reference depth (a/d < 1), the current density is nearly uniform when excited from one side but zero at the center when excited from both sides.

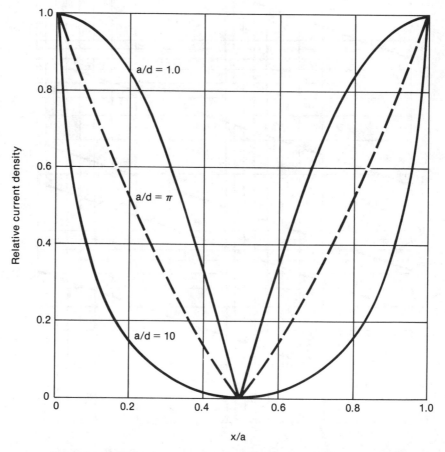

Fig. 3.6. Relative current-density distribution through a metal sheet heated from both sides as a function of the ratio of sheet thickness to reference depth (a/d) (Ref 38)

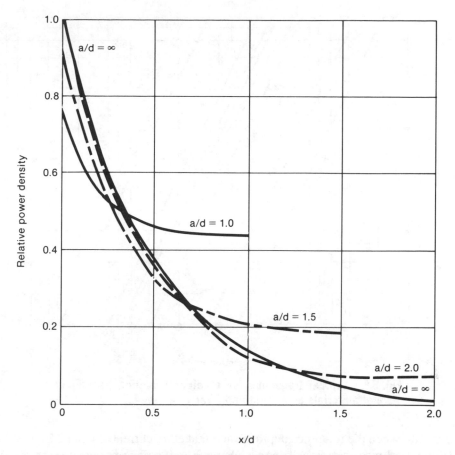

Fig. 3.7. Relative current-density distribution through a metal sheet heated from one side only as a function of the ratio of sheet thickness to reference depth (a/d) (Ref 38)

Only when the material is several reference depths thick does the power distribution approach the exponential shape of an infinitely thick material.

A very important result of the skin effect is evidenced in heating efficiency. Heating efficiency is the percentage of the energy put through the coil that is transferred to the workpiece by induction. It can be shown that if the ratio of workpiece diameter to reference depth for a round bar drops below about 4 to 1, the heating efficiency decreases. The critical frequency is defined as the frequency at which the ratio of workpiece diameter to reference depth is about 4 to 1 for round bars; if a sheet is heated from both sides, the critical ratio of thickness to reference depth is 2 to 1. Figure 3.8 shows the critical frequency as a function of diameter for round bars, and Fig. 3.9 shows the efficiency of heating as a function of this critical frequency. In the latter figure, it can be seen that efficiency increases relatively little with large increases in frequency above the critical frequency. Below the critical frequency, efficiency drops rapidly because less current is induced due to current cancellation. Current cancellation becomes

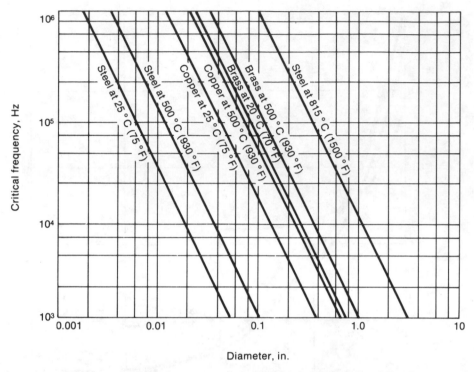

**Fig. 3.8.   Critical frequency for efficient induction heating of
several materials as a function of bar size (Ref 42)**

significant when the reference depth is such that eddy currents induced from either
side of a workpiece "impinge" upon each other and, being of opposite sign, cancel
each other.

For through-heating, a frequency close to the critical frequency should be
chosen. In contrast, for shallow heating of a large workpiece, a high frequency
should be selected. In this instance, there is no concern about critical frequency,
because the diameter will be many times the skin depth. For steels, surface heating
for hardening purposes is one of the most common applications of the induction
process. In general, the surface is heated to a temperature above the upper critical
temperature and then quenched. More will be said about frequency selection for
hardening of steel in Chapter 5.

Effect of Frequency on Mechanical Forces

Another reason for proper planning and choice of frequency is that mechanical
forces act on the load in an induction heating system. The forces involved are the
same ones present in motors and relays — that is, between two parallel, current-
carrying conductors, there is a force which is attractive (currents flowing in
opposite directions) or repulsive (currents flowing in the same direction). The two
current-carrying conductors in an induction heating application are the coil and the
workpiece itself. The forces are important because they tend to change the position

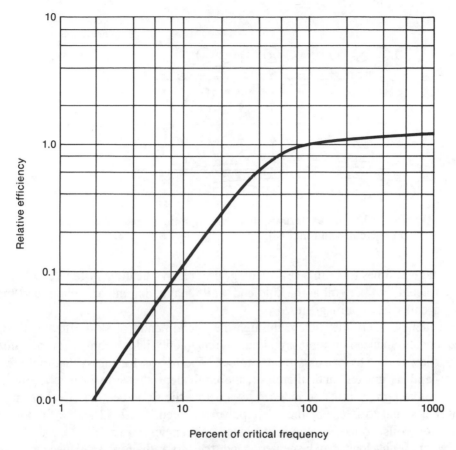

Fig. 3.9. Relationship between relative efficiency and critical frequency (Ref 42)

of the load in the coil if the load is not centered and the forces are not balanced, or to distort the coil if it is not adequately supported.

Consider first the forces on a *nonmagnetic* cylindrical material in a solenoidal coil. In this case, the forces are proportional to the product of the currents in the coil and workpiece and to the length of the workpiece, and are inversely proportional to the distance between the coil and workpiece. If the cylinder is centered in the coil, the forces within the workpiece are inward, perpendicular to the surface. The forces at the ends are not necessarily perpendicular to the surface but are balanced. Therefore, there is no net force on the load. If the load is not centered (Fig. 3.10), the forces due to the fringing at the end cause the bar to have a net force toward one end. In heating of nonmagnetic loads, because of this effect, it is often necessary to provide means for holding the load in the coil.

The net force on a *magnetic* load may be the reverse of that on a nonmagnetic load. Here, there is an additional attractive force between the magnetic load and the coil. When this force is greater than the repulsive force due to induced currents, the load will be attracted into the coil. These forces can be used to

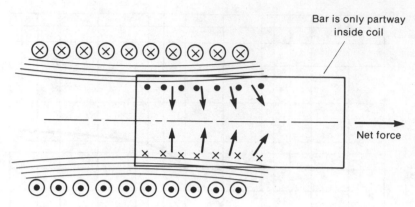

Fig. 3.10.   Forces (arrows) on a nonmagnetic bar partially inside an induction coil (Ref 43)

advantage at times to form a self-feeding, progressive heating system. Steel bars are attracted into the coil at the cold end while magnetic, and are repulsed at the heated end when nonmagnetic.

Frequency enters the picture through the phenomenon of skin effect. At low frequencies, induced currents are larger and magnetic fields stronger for the same power level, thereby causing greater forces. The reason for this is that the reference depth is greater, and, therefore, the equivalent sleeve resistance is lower. Thus, more current must be induced to generate the same heating effect because power is equal to I^2R. For the same power induced in a load and with similar induction coils, forces are lower at higher frequencies.

Even though a load may be centered and, therefore, the forces balanced and not tending to eject the load from the coil, there are forces repelling the induction coil from the load in a radial direction. Thus, coils require strong supports at low frequencies. The force between the heating coil and the workpiece manifests itself particularly with pancake heating coils, which are used for heating flat surfaces from one side. Unless the coil is carefully braced, it will move away from the workpiece each time the power is applied. This not only disturbs the uniformity of the air gap between the coil and the workpiece, altering the heating pattern, but also harms the coil itself. Repeated flexing gradually weakens the coil and may cause it to break.

Therefore, in choosing a frequency, careful estimates of force should be made, and experiments to verify them may be needed. If forces appear to be too great to be tolerated, it may be necessary to go to a higher frequency.

DETERMINATION OF POWER REQUIREMENTS

The amount of power that will be needed for a given application is also one of the major parameters to be determined. If the workpiece is of regular shape and is to be through-heated, the calculation is straightforward. However, if it is to be selectively heated so that the remainder of the workpiece is a heat sink for the generated heat, calculation of the power needed may be difficult. Such a calculation is not considered in the present discussion, although rules of thumb for surface heating for purposes of surface hardening of steel will be given in Chapter 5.

Consider a load (workpiece) that is to be heated throughout the entire volume. For such through-heating applications, the power should be kept relatively low to allow heat conduction from the outer layers (which are heated more rapidly by higher current densities) to the inner layers. There will always be a temperature gradient, but this can be minimized by careful selection of induction heating parameters. Neglecting the temperature gradient, the absorbed power depends on the required temperature rise, ΔT; the total weight to be heated per unit time, W; and the specific heat of the material, c. The power to be supplied to the load (specific heat power), P_1, is then equal to $Wc\Delta T$.

To determine the total input power needed from the power source supplying the ac current to the induction coil, the power lost from the workpiece due to radiation and the power lost in the coil itself by I^2R effects must be added to P_1. Heat loss by convection to the surrounding atmosphere (if not in a vacuum) should also be considered. This loss depends on the thermal characteristics of the atmosphere, the velocity of the atmosphere (free or forced convection), and the size of the workpiece. For typical rapid heating applications, the convective heat loss is usually small and is neglected in calculations of power requirements.

Radiation power losses are calculated by use of the expression:

$$P_2 = Ae\sigma(T_2^4 - T_1^4)$$

where e is the emissivity of the workpiece; σ is the Stefan-Boltzmann constant; T_1 and T_2 are the workpiece and ambient temperatures (in K), respectively; and A is the surface area of the workpiece. The radiation power loss varies greatly during the heating cycle because the emissivity changes with workpiece surface condition (e.g., it becomes oxidized). Typical emissivities are 0.1 to 0.2 for aluminum at 200 to 595 °C (390 to 1100 °F) and 0.80 for oxidized steel.

Power lost in the work coil, P_3, is given by:

$$P_3 = (P_1 + P_2)\left(\frac{1}{\text{eff}} - 1\right)$$

where P_1 is the power required to heat the workpiece, P_2 is the radiation loss, and eff is the coupling efficiency of the coil and workpiece. This quantity depends on coil design and the size of the air gap between the inductor and workpiece, among other factors. Average coupling efficiencies for closely coupled loads are given in Table 3.1. The total power to be supplied by the induction heating generator is the sum of P_1, P_2, and P_3.

Table 3.1. Average coupling efficiencies for closely coupled loads (Ref 44)

Type of coil	Average coupling efficiency(a) for:			
	Magnetic steel below Curie temperature	Steel above Curie temperature; stainless steel	Brass; titanium; aluminum; bronze	Copper
Helical:				
Around workpiece 0.90		0.65	0.50	0.30
Internal 0.70		0.40	0.30	0.20
One turn:				
Around workpiece 0.85		0.60	0.45	0.25
Internal 0.65		0.35	0.25	0.15
Hairpin 0.85		0.60	0.45	0.25
Pancake 0.70		0.40	0.30	0.20

(a) For loads with approximately 0.32 cm (0.125 in.) between coil and workpiece.

Sample calculations for heating of steel for hardening purposes are given in Chapter 5.

Chapter 4

Equipment and Coil Design for Induction Heating

In Chapter 3, the basic principles underlying induction heating were described. The basic component of the system is an induction coil (or inductor) which is energized by an alternating current. Proper selection of the frequency of this current is determined by the type of application (surface or through-heating), part size, and the electrical and magnetic properties of the workpiece. Because of the wide variety of applications, it is not surprising that a wide range of frequencies are encountered in induction heating applications. Hence, different types of equipment are needed to generate alternating currents in the different frequency ranges. In this chapter, the various common types of induction generator equipment are described. In addition, coil designs for various applications are discussed, and means by which the coil and generator are "matched" are summarized.

POWER SOURCES

Power sources for induction heating have changed over the years. In the high-growth years of induction heating (about 1945 to 1965), the primary types of power sources were systems using line-frequency current (60 Hz), motor-generator systems, and vacuum-tube systems. In the late 1960's, solid-state systems, which have generally replaced motor-generator systems in new installations, were introduced. Table 4.1 gives frequency and power ranges, approximate efficiencies, and other features of the four major power sources. Choice of a power source depends primarily on frequency, although there is some overlap among the frequency capabilities of the different sources (Fig. 4.1).

Line-Frequency Systems

The major advantage of this type of system is the fact that there is no frequency conversion. Power losses are therefore reduced, and the system is greatly simplified by the absence of either rotating components or complex electronic equipment. Maintenance is also reduced, and switchgear, power-factor correction

Table 4.1. Characteristics of the four major power sources for induction heating

Power source	Frequency range	Power range	Efficiency, %	Features
Line frequency . . .	60 Hz	100 kW to 100 MW	90 to 95	High efficiency; low cost; no complex equipment; deep current penetration
Motor-generator . .	500 Hz to 10 kHz	10 kW to 1 MW	75 to 85	Low sensitivity to ambient heat; low sensitivity to line surges; fixed frequency; low maintenance cost; spares not needed
Solid state	180 Hz to 50 kHz	1 kW to 2 MW	75 to 95	No standby current; high efficiency; no moving parts; needs protection outdoors; no warm-up time; impedance matches changing loads
Vacuum tube	50 kHz to 10 MHz	1 kW to 500 kW	50 to 75	Shallow heating depth; localized heating; highest cost; impedance matches changing loads; lowest efficiency

capacitors, control circuits, etc., can be selected from standard ranges. Additional problems, such as radiation losses, stray-field heating, and unwanted high-frequency induced currents, are negligible. Cost of the basic equipment is also generally less than that of medium- or radio-frequency units. The major parts of a typical line-frequency system are shown in Fig. 4.2.

From the application viewpoint, the main use of line-frequency or 60-Hz sources is for through-heating. Because most application requirements in through-heating include maximum temperature uniformity, these frequencies are more applicable than the higher ones. This was seen in Chapter 3 in which it was noted that the greatest current penetration results from the lowest frequency. Induction melting furnaces also use line frequencies extensively.

The fundamental disadvantage of low-frequency heating, especially at 60 Hz, is the decrease in power input for a given workpiece size and coil magnetic intensity. This disadvantage is often wholly compensated for by the increased temperature uniformity, enabling a minimum time for heat to soak in from the surface. Surface power-density figures are usually lower for through-heating, but workpieces are much larger. Round billets often measure 12.5 to 35 cm (5 to 14 in.) in diameter; slabs can be up to 30 by 200 cm (12 by 80 in.) in cross section with lengths ranging up to 10 m (400 in.). Therefore, coil and system power requirements are usually high. A single coil may take several megawatts of power, and systems requiring over 30 MW have been installed.

Due to the high power levels involved, balanced three-phase systems are nearly always used.* In general, most nonferrous through-heating is carried out with

---

*Line power is usually single-phase or three-phase alternating current. Single-phase current is produced by a transformer with one primary and one secondary. Current is generated in the

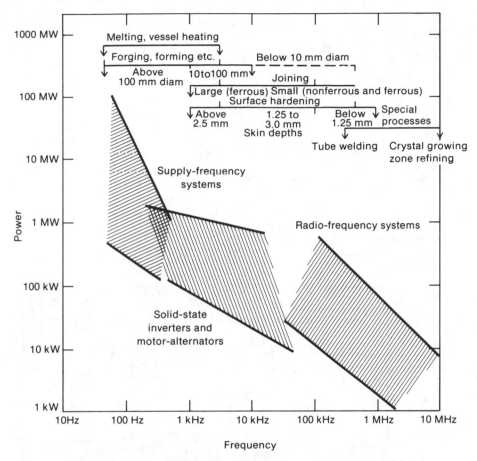

Fig. 4.1. Power and frequency ranges for common types of induction heating generators (Ref 40)

single-coil inductors. These may consist of a three-phase coil, the parts of which are energized by the different phases. Such arrangements yield a reasonably balanced load, provided that the coil is filled with billets. If the coil is only partly filled, one phase becomes out of balance. For this reason, it is good practice to avoid a neutral wire in the supply line, which could carry an excessive current. Sometimes, the higher temperatures required for heating steel above the Curie temperature (at which $\mu = 1$) are obtained with medium frequencies, and, in these dual-frequency systems, the number of 60-Hz coils is such that each can be energized separately by a single phase. This enables a well-balanced three-phase load to be presented to the supply. Thus, three small single-phase systems can be loaded on a three-phase supply.

---

secondary by induction methods. Three-phase power is produced by three sets of windings. In each set, there is a primary and a secondary. The currents in each set are out of phase with those in the other sets by 120°. That is to say, the sinusoidal variation of current with time is displaced one-third of a cycle for each phase.

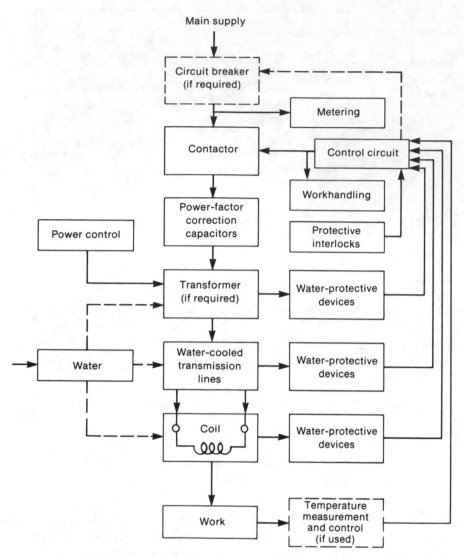

Fig. 4.2. Basic components of a line-frequency induction generator system (Ref 40)

Motor-Generator Systems

As their name implies, motor-generator systems use motors with copper windings in order to take advantage of the principles of electromagnetic induction to convert line-frequency current into higher-frequency current. For many years, the only source of medium-frequency power was the medium-frequency motor-generator; these devices have been almost totally superseded by their solid-state rivals. They are still widely used as induction heating supplies.

While solid-state converters have made substantial inroads into the medium-frequency market on grounds of higher efficiency, careful consideration should be given before the motor-generator system is rejected. The prospective customer should take into account its proven reliability with trouble-free running for many years and minimum upkeep. The potential user should also consider the complexity of the solid-state device and the kind of technical skill needed for its maintenance, compared with the occasional routine attention to the motor-generator bearings by relatively unskilled personnel and the ready availability of rewind facilities should a rare major fault develop. Some value should also be placed on the accumulation of over 40 years of experience in the use of these generators for various heating applications. Most of the problems were met and solved long ago.

Frequencies near supply frequency, which are multiples of supply frequency, e.g., 180 Hz, can be produced with motor-generators using a standard induction motor driving an ac generator. In this case, the output windings are wrapped on the rotor. Because of this, these windings can be subjected to large centrifugal forces at high speeds.

A subclass of motor-generator is the inductor-alternator, which has no windings on the rotor. In this instance, both the excitation and output windings are on the stator; therefore, there are no centrifugal forces acting on the output windings, and the rotor can be an extremely robust package of silicon steel punchings. Such designs permit higher frequencies and power capacity in the basic motor-generator class of equipment. Motor-generator power sources come in many power, voltage, and frequency ranges. Table 4.2 gives one manufacturer's ranges which are typical for the industry.

Solid-State Systems

For production operations requiring medium-frequency induction generators, attention should be given to the alternative of either a motor-generator or a solid-state power supply. The primary intent of this section is not to present an in-depth comparison of motor-generators with solid-state devices. Nevertheless, it is appropriate to list the major points requiring evaluation.

Motor-generators:
- Were the prime medium-frequency converters for induction heating for over 35 years.
- Possess a high level of production reliability resulting from many years of production experience.
- Are produced at present by only two or three manufacturers. Delivery schedules are long as a result of limited production capacities. Very few motor-generators have been delivered since 1980.
- The low-slip motor characteristics require high current starting capability from the line power supply.
- Consume 15 to 20% of maximum power when running in a standby condition, i.e., not heating.
- Have rotational speeds on the order of 3600 rpm with closely fitting components requiring precision manufacturing tolerances.

Table 4.2. Standard continuous ratings of motor-generators from one manufacturer (Ref 40)

Output, kW	Frequency, kHz	Voltage, V	Motor rating, kW
350	1	1200/600	400
350	3	1100	400
350	4	1100/500	400
300	1, 3, or 4	1200/600	332/343
225	8.8	1000/500	276
180	10.4	1000/500	224
200	10.4	1000/500	261
200	1	800/400	228
200	3 or 4	800/400	231
220	4	800/400	254
150	8.8	1000/500	187
120	10.4	1000/500	153
150	3	800/400	172
110	8.8	800/400	134
85	10.4	800/400	110
100	3	800/400	116
75	8.8	800/400	93
60	10.4	800/400	78
80	3	800/400	93
45	10.4	800/400	59
50	3	800/400	60
30	10.4	800/400	41

- Require limited maintenance, except for important lubrication schedules. However, when major problems occur, expensive and lengthy repairs may be involved.
- Suffer continual price escalations resulting from inability to reduce construction costs and from the limited supply.

Solid-state generators:

- Are specified for many new induction heating installations.
- Are produced by numerous manufacturers who are usually directly associated with the individual induction heating system supplier.
- Are characterized by favorable electrical energy factors with no starting current requirements, higher conversion efficiencies, and virtually no power consumption when not heating for production.
- Are lower in cost than motor-generators.
- Do not require extended periods for repair.
- Can be delivered on reasonable schedules dependent on the associated equipment for the induction heating system.
- Are priced competitively because of the large number of manufacturers. However, the lowest cost selection may involve considerable evaluation.

It is not possible to cover thoroughly the design details of solid-state systems in this chapter. Peschel (Ref 45) has discussed them in greater depth. Only some brief descriptions of several basic circuits will be given. It is emphasized, however, that solid-state systems have replaced motor-generator systems in many new and existing installations. When motor-generators fail, they are usually replaced by solid-state power supplies.

DC Power Supplies. With the exception of the ac-to-ac inverter, which is not common and will not be discussed here, all of the medium-frequency solid-state generators commercially available for induction heating applications are of the ac-to-dc-to-ac type. They convert ac line power to dc and then back to ac of a given frequency. Most of them have essentially the same ac-to-dc power conversion circuit as shown in Fig. 4.3. Several inverter designs utilize SCR's (silicon controlled rectifiers) in the place of diodes in the three-phase rectified bridge for variable dc power control for start-up, shutdown, and temperature control.

DC-to-AC Power Conversion. The most significant difference in commercially available solid-state generators is with the dc-to-ac power conversion circuit. Typical inverters for this purpose include the (1) half-bridge, series load, (2) reactor, (3) full bridge, (4) current feed, and (5) charge-discharge types. In all cases, the "turn off" times of SCR's limit the frequencies of solid-state power supplies to a maximum of approximately 50 kHz.

The half-bridge, series load inverter is the simplest conceptual design. In this case (Fig. 4.4), two SCR's are activated alternately by a pulse generator, setting

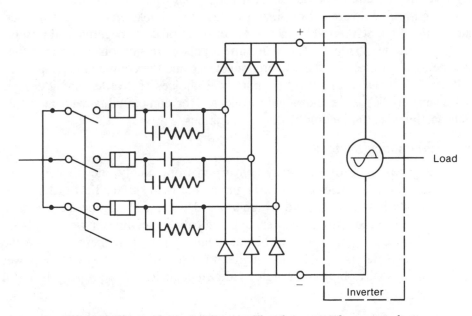

Fig. 4.3. Three-phase bridge rectifier for ac-to-dc conversion (Ref 45)

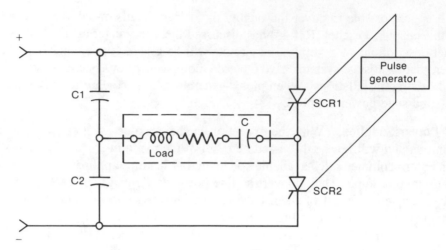

Fig. 4.4. Half-bridge, series load inverter (Ref 45)

up an alternating current through a simple inductor-capacitor (L-C) circuit. The inductor is the induction coil itself.

The reactor inverter circuit (Fig. 4.5) was the first commercial SCR power circuit to be successfully adapted for 1000- and 3000-Hz induction heating and melting applications. The basic design uses the SCR as a switch to charge and discharge a capacitor through an inductor, thus providing a half-cycle of the current pulse to the resonant tank circuit* of the load. Alternate firing of the negative inverter circuit produces the negative half-cycle of pulsed energy.

The full-bridge inverter (Fig. 4.6) is a classic textbook circuit requiring alternate firing of the SCR's (which act as switches) to produce negative and positive energy pulses into the load. The SCR current pulses are sinusoidal, with the time base being determined by the air core inductance and the commutation capacitor.

It can be concluded that there are numerous choices of inverter circuit designs and characteristics to be thoroughly evaluated for compatibility with industrial plant facilities and induction heating applications.

Radio-Frequency Systems

The major purpose of frequency conversion from supply-line frequencies of 60 Hz to radio frequencies of 100 kHz and higher is to surface heat materials to shallow depths in short times and at high power densities. Most generators in these instances are used for surface hardening and localized heating for welding, brazing, etc. A second purpose is to obtain greater power densities. Because the equivalent cylinder is thinner and higher in resistance, higher surface-power densities are usually obtained at higher radio frequencies. This property is useful

---

*A tank circuit is an L-C circuit which has been tuned (i.e., values of L and C have been selected to match the impedance of the two components) to resonance and thus appears to be a pure resistance to the power source.

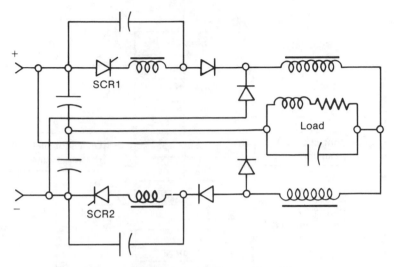

Fig. 4.5. Reactor inverter (Ref 45)

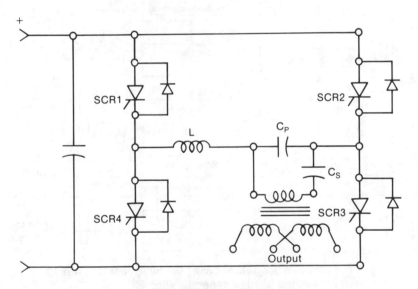

Fig. 4.6. Full-bridge inverter (Ref 45)

when the coupling between coil and workpiece has to be very poor — e.g., when the workpiece is surrounded by a protective atmosphere and its container. The effective coil-power factor is then very low, and considerable tank-circuit volt-amperes are necessary to supply the number of kilowatts required.

All radio-frequency generators consist of several conversion stages, as shown in Fig. 4.7. Figure 4.8 is a schematic of a basic oscillator circuit* with the

---

*An oscillator circuit is a vacuum-tube radio-frequency generator. The frequency developed is a function of the values of L, C, and R in the tank circuit, which forms a basic portion of the generator.

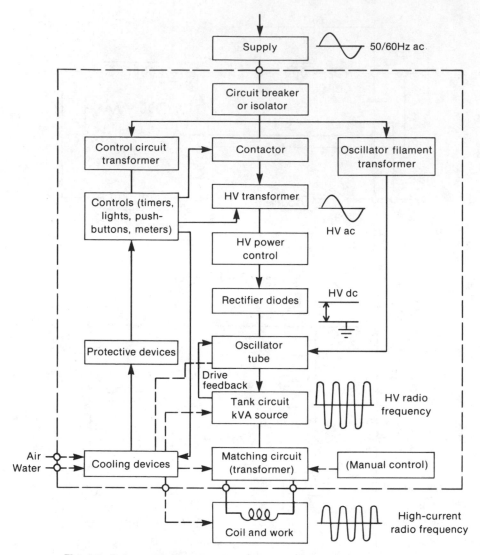

Fig. 4.7. Schematic diagram showing the components of a radio-frequency induction heating generator (Ref 40)

workpiece coil in the most common position, in series with the tank coil. At the output end, the coil and workpiece present a power factor of 20% or less to the output terminals. This factor is usually inverted and the value expressed as a Q factor.* Most radio-frequency loaded work coils have Q values between 5 and

---

*The Q factor is defined in general as Q = 2π (total energy stored/energy dissipated per cycle). For electrical circuits, it is often given approximately as Q = 2πfL/R, where f is frequency. The Q factor of a given tank circuit with a given load coil and workpiece thus depends on the frequency. The power factor is also equal to the cosine of the phase angle between the current and voltage in an electric circuit. For both a pure inductor and a pure capacitance, the phase angle is 90°, resulting in a power factor of 0 and a Q factor of infinity.

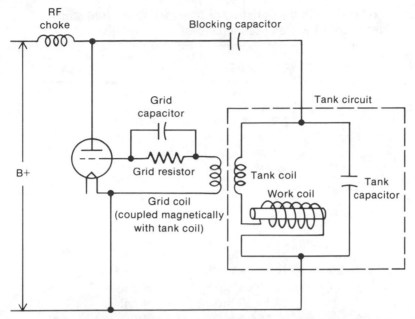

Fig. 4.8. Basic high-frequency oscillator circuit (Ref 46)

15 for ferrous loads and between 10 and 25 for nonferrous loads, occasionally rising to 100 or more for very loosely coupled nonferrous loads. For efficient induction heating, small Q values are desirable.

Radio-frequency systems are manufactured in a range of power levels and frequencies. Table 4.3 lists the standard line of one manufacturer.

INDUCTION COIL DESIGN

Although choosing the proper frequency is possibly the most important step in designing an induction heating system, there is often some flexibility in the frequency range that is adequate. Designing an induction coil, however, is very specific to a given application and can be considered the most important aspect of the engineering of the system. There are several major functions that an induction coil must perform to make a job successful:

- Induce current in the load so that the proper heating pattern is obtained.
- Accomplish the proper heating pattern with as great an efficiency as possible.
- Provide an impedance match to the generator so that adequate power can be transferred to the load.
- Have a geometry that will accomplish the above three major functions and permit easy loading and unloading of the part being heated.

Heating Pattern Control

The single most important aspect of an induction heating application is the development of the desired heating pattern. Other considerations, such as how a coil is matched with the generator, are of secondary importance.

Table 4.3. Characteristics of standard radio-frequency generators from one manufacturer

Output, kW	Frequency	Input V	kVA
1	400 kHz	115	2.3
2.5	400 kHz	230, 460	5.5
5	250 to 450 kHz	230, 460	12.5
7.5	250 to 450 kHz	230, 460	17
10	250 to 450 kHz	230, 460	23.5
15	180 to 450 kHz	230, 460	35
20	180 to 450 kHz	230, 460	45
25	180 to 400 kHz	230, 460	55
30	180 to 400 kHz	230, 460	70
40	180 to 400 kHz	230, 460	92
50	180 to 400 kHz	230, 460	120
75	180 to 400 kHz	230, 460	165
100	180 to 400 kHz	230, 460	210
2.5	2.5 to 8 MHz	230, 460	5.5
5	2.5 to 8 MHz	230, 460	12.5
7.5	2.5 to 8 MHz	230, 460	17
10	2.5 to 8 MHz	230, 460	23.5
15	2.5 to 8 MHz	230, 460	35
20	2.5 to 8 MHz	230, 460	45
30	2.5 to 8 MHz	230, 460	70
50	2.5 to 5 MHz	230, 460	120

For simple geometries, the desired heating pattern can often be obtained easily. The simplest case is the classic example of heating a section of a long cylinder by a simple solenoid coil. This was illustrated in Fig. 3.3. If the coil is closely coupled (in which case the workpiece diameter is a large percentage of the coil diameter), the heating pattern will be of approximately the same length as that of the coil and will be relatively uniform. If the load is still essentially cylindrical but slightly modified, such as in a crankshaft bearing where heat is desired in the fillets and entry and removal of the shaft must be made easy, some rather large alterations must be made to a simple solenoid coil. If the part contains an unusual protruding shape, an unusual shape must also be built into the induction coil. Much induction coil design depends to a large extent on experience. Only general guidelines can be pointed out, mostly from actual examples. Therefore, some basic induction coil shapes and the types of loads for which they are applicable will be illustrated.

Variations of the solenoid coil in which its extent is greater or less than that of the workpiece are shown in Fig. 4.9. This figure also illustrates solenoid coils used for heating the internal surfaces of tubular parts; internal coils, although much less efficient than external ones (Table 3.1), should not be overlooked in potential applications. It can be seen in the figure that coil design has a marked effect on heating pattern. Modifications of these patterns, if a more uniform pattern is desired, can be obtained by contouring the coil or varying the pitch of the coil as suggested in Fig. 4.10.

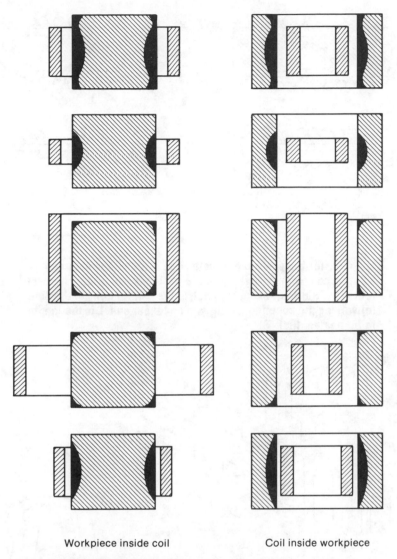

Workpiece inside coil   Coil inside workpiece

Fig. 4.9. Effect of relative proportions of workpiece and coil on heating patterns in round parts (Ref 47)

A very large number of complex-geometry coils can be designed. Their shapes and applications are limited only by the ingenuity of the designer. Examples of such coils (and the resulting heating patterns) are shown in Fig. 4.11 to 4.14.

The channel coil (shown in Fig. 4.11) is a class of coil that is widely used; it is probably the most common type after the solenoid coil. It is used to heat components from one end or one side only and parts that can or must be moved through the coil in a continuous process. This is possible inasmuch as the coil does not surround the part. The channel coil is frequently used for continuous ("progressive") heating of bars for forging or as a heating device for brazing. Other types of coils and processing setups (e.g., Fig. 4.12c and d) also allow for progressive heating.

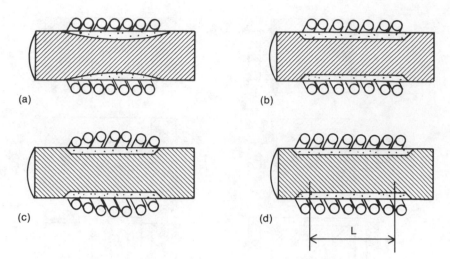

Fig. 4.10. (a) Uneven heating pattern in a round bar obtained by using a coil with an even pitch, a problem which can be corrected by: (b) increasing the pitch of the central turns of the coil, (c) varying the coupling, or (d) using a longer coil. L is the length to be heated. (Ref 48)

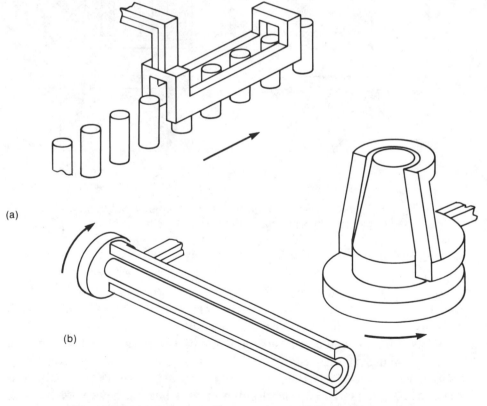

Fig. 4.11. Typical designs of (a) a channel (slot) induction coil and (b) single-shot, nonencircling coils (Ref 49)

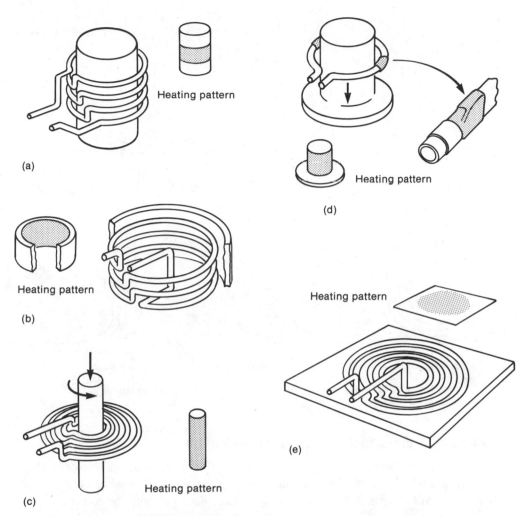

Fig. 4.12. Typical coil designs for surface hardening via high-frequency induction (Ref 50)

A popular alternative to the solenoid coil for heating round parts is the non-encircling coil (Fig. 4.11b). With this coil, the induced currents flow longitudinally rather than circumferentially. The part must be rotated during operation to ensure uniform heating of the entire surface. This coil design has become popular in recent years for surface hardening of various types of shafts.

Another commonly used coil geometry is the pancake configuration shown in Fig. 4.12(e) and 4.13(d). This type of coil is used to heat materials from one side only, but can be modified for parts which are contoured or curved. Other specially contoured coils are illustrated in Fig. 4.13 and 4.14.

In almost all instances, induction heating coils are constructed from copper tubing. Copper is selected because of its low electrical resistance and is used in tubular form so that it may be water cooled to prevent overheating during operation.

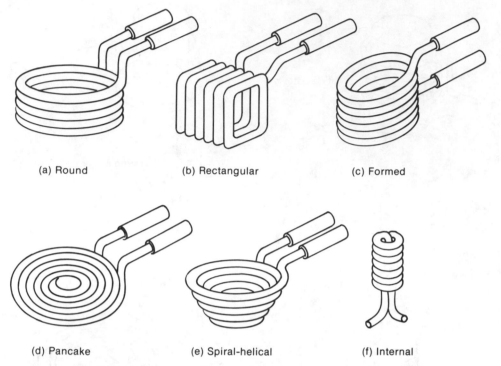

(a) Round   (b) Rectangular   (c) Formed

(d) Pancake   (e) Spiral-helical   (f) Internal

Fig. 4.13. Multiturn coils for various induction heating applications (Ref 51)

Matching of the Induction Coil to the Generator

Of the various aspects of coil design, matching the coil to the generator is also very important. Even though a heating pattern for the workpiece has been specified and a coil designed that will produce this desired pattern, the coil must still be matched to the generator to achieve the heating desired in the required time. If the full power rating of a generator is to be realized, it is essential that the electrical conditions of the system be such that full rated current is drawn from the power supply at rated voltage and that the generator see a power factor as close as possible to unity. This is called impedance matching. Impedance matching involves two variations of the same basic problem: matching to a fixed-frequency source and matching to a variable-frequency source. Typical matching methods include changing the number of induction coil turns and the use of a matching transformer.

Fixed-Frequency Source Matching. The two most common fixed-frequency sources are line-frequency sources and motor-generator systems. In these two types of systems, the load-coil voltage is set by the operator, but the load current depends on the load impedance. Such systems require a power factor of unity and the proper load impedance to draw rated power from the source. The impedance of the load seen by the generator is determined by the number of turns of the

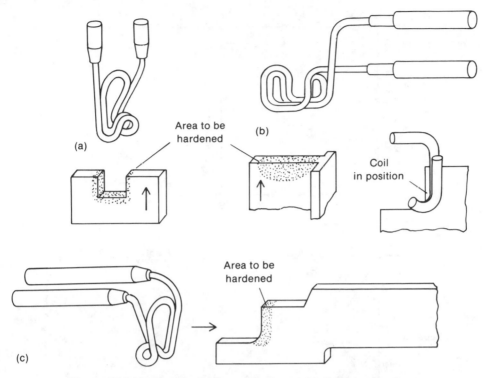

Area to be hardened

(a)

(b)

Coil in position

Area to be hardened

(c)

Fig. 4.14. Odd-shape induction coils for specialized applications involving localized heating (Ref 51)

induction coil around the load or the number of turns of the coil as modified by the ratio of an output transformer. Calculating the number of turns necessary to match a given source with a given rated voltage and current is a rather simple process requiring knowledge only of the geometry of the coil and load and the electrical properties of the load. The calculation of the number of turns of the coil is described in various papers (Ref 52 to 54). The calculation can be understood broadly by referring to Fig. 4.15. Figure 4.15(a) shows the circuit of a fixed-frequency system with a multiple-turn induction coil and a cylindrical load represented as a single-turn coil. The load-coil circuit is tuned to achieve a unity power factor by a parallel capacitor. It is desired to calculate the number of turns in the load coil that will present a proper match to the generator and also the capacitor value that will bring the power factor to unity for a given load in the coil.* Figure 4.15(b) shows the equivalent tank (capacitor-inductor) circuit. R_A and X_A are the transmission-line resistance and reactance which generally can be considered to be zero if the coil is within a few feet of the capacitor. R_P, and X_P and X_O, are the resistance of the coil and the reactance of the coil and the air gap,

---

*The concept behind this procedure is discussed in Appendix C; a computer program for carrying out the required calculations is given and explained in Appendix D.

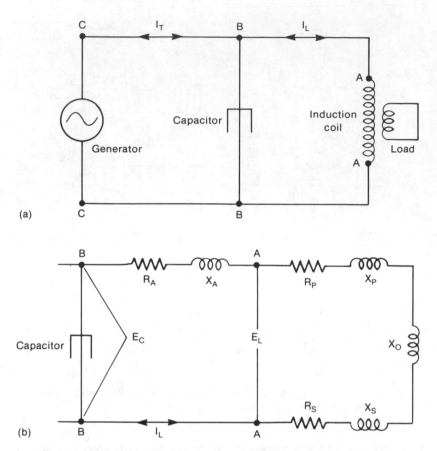

Fig. 4.15. (a) Schematic diagram of a typical induction heating circuit using a motor-generator set and (b) equivalent electrical circuit from capacitor through load (Ref 52)

respectively. R_S and X_S are the resistance and reactance of the load as reflected into the induction coil circuit. These latter five values are calculated from the geometry of the load and coil, and, from these five calculated parameters, the turns in the induction coil, the capacitor value, and the current in the coil can be calculated to match the given rated voltage, current, and power of the generator.

The above discussion assumes, among other things, that the workpiece is a solid bar which is long relative to its diameter and that the reference depth is much smaller than the bar diameter. There are several factors that can be applied in the basic calculation to correct for variations from this ideal case. Figures 4.16 and 4.17 show two of them: the so-called workpiece resistance factor as a function of the ratio of bar diameter to reference depth, a/d; and the workpiece shortness factor, which is a function of coil and air-gap dimensions. Note that the "knee" in Fig. 4.16 occurs at about the same critical ratio ($a/d = 4:1$) required for efficient induction heating, which was discussed in Chapter 3. R_S in Fig. 4.15(b)

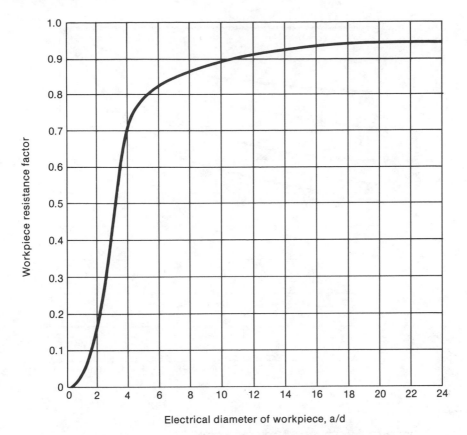

Electrical diameter of workpiece, a/d

**Fig. 4.16. Workpiece resistance correction factor used for calcu-
lating workpiece resistance as reflected in induction coil circuit
for solid round bars (Ref 43)**

can then be calculated from the equivalent cylinder resistance multiplied by the
turns ratio. This quantity is then multiplied by the above two correction factors.
A slightly different workpiece resistance correction factor (Fig. 4.18) is employed
if the load is a thin-wall tube rather than a solid bar.

Solid-State System Matching.   The solid-state converter is a hybrid between the
fixed-frequency source and the radio-frequency source whose frequency varies
with load impedance. Most solid-state systems have a "controlled" frequency — that is, changes in load impedance are sensed through the resonant circuit voltage, and the frequency is varied by firing the SCR's at the desired rate. This
source frequency is not necessarily the resonant frequency of the load-tank circuit.
However, for load-matching purposes, the calculations can be made essentially
as they are made for the fixed-frequency source: the number of turns in the coil
and the capacitor values are estimated from the geometry of the situation and the
source rating. During heating, however, the load will not necessarily draw the full

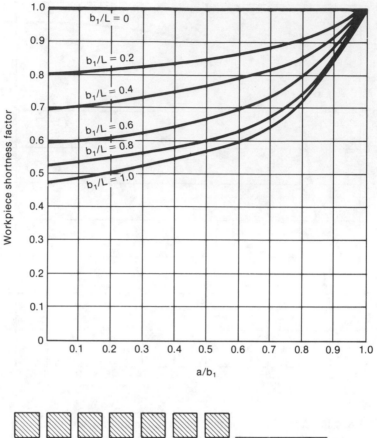

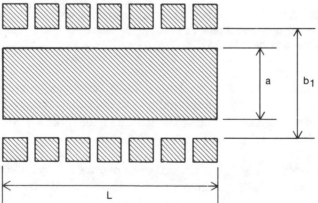

Fig. 4.17. Workpiece shortness correction factor used for calculating workpiece resistance as reflected in induction coil circuit for solid round bars (Ref 43)

rated power of the generator when workpiece properties undergo drastic changes (such as when the temperature of a steel part passes through the Curie point).

Variable-Frequency Source Matching.   Matching to a radio-frequency generator is much easier than matching to a fixed-frequency source. In essence, the fre-

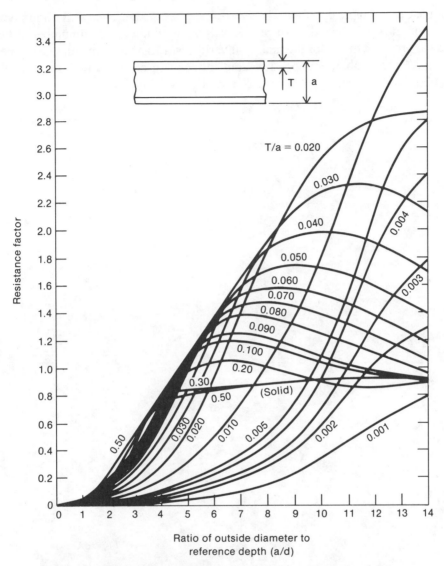

Fig. 4.18. Workpiece resistance factor used for calculating work-piece resistance as reflected in induction coil circuit for round, hollow loads (Ref 43)

quency of the oscillator circuit is controlled by the impedance values in the tank circuit, a part of which is formed by the induction coil and the workpiece. The tank circuit is always at resonance and therefore always appears to be a pure resistance (power factor of unity) in the oscillator circuit. Thus, changes in the magnetic and electrical properties of the load can be tolerated within fairly wide limits while still drawing a large fraction of the power from the generator.

In principle, the number of turns needed for a load coil can be calculated as discussed in the preceding section. However, this is seldom done. In the first place, loads for radio-frequency generators often are small and more geometrically complex than simple cylinders. In the second place, experience will often give an operator enough knowledge for proper estimation of the required number of turns and the proper settings for the tank-coil variables. Most tank coils in these generators have changeable, or tapped, output fittings that attach to the load coil, thus enabling coils of varying design to be used effectively.

Chapter 5

Application of Induction Heating to Surface and Through-Hardening of Steel

In Chapters 2 and 3, the fundamental principles underlying heat treatment of steel and induction heating were briefly summarized. Chapter 2 described the austenitizing, hardening, and tempering processes that form the basis for final heat treatment of a large number of steels. In Chapter 3, the method by which eddy currents are induced in metals, thereby leading to heating, were discussed; this discussion dealt with such important considerations as choice of frequency and power. In the present (and following) chapters, the application of these concepts to heat treatment of steel by induction methods is detailed. This chapter will concentrate on hardening by induction.

METALLURGICAL CONSIDERATIONS IN INDUCTION HARDENING

As mentioned in Chapter 2, the primary means of hardening steel through heat treatment processes consists of the steps of austenitizing, quenching, and tempering. The first two of these steps serve to harden the steel. As with parts hardened by conventional heat treating methods, the hardness and hardness profile of induction austenitized parts depend solely on the chemical composition of the steel and the quenching medium. Therefore, the primary metallurgical question that arises in induction hardening relates to austenitizing.

As with austenitizing by conventional furnace techniques, time and temperature are the two critical parameters that must be controlled. To completely austenitize a steel, a certain amount of time at or above the A_3 temperature is required. For rapid heating processes such as induction, literature data on austenitizing are of two types: (1) austenitizing temperatures for continuous, rapid heating (data which have greatest application to surface hardening); and (2) "isothermal" time-temperature austenitization data (which have more application to through-hardening processes). The first of these is for the lower and upper critical temperatures that pertain under continuous rapid heating. In other words, the Ac_1 and Ac_3 temperatures are given as functions of heating rate. Because the only "soaking" time available for phase transformation in these cases is that time after which the equilibrium critical temperature (A_1 or A_3) is exceeded, the continuous-heating critical temperatures (Ac_1, Ac_3) are always *above* the equilibrium ones. As might be expected, this difference increases with heating rate—an effect shown in Fig. 5.1 for the Ac_3 temperature for 1042 carbon steel. Here, the Ac_3 temperature is the one at which it has been estimated that austenite reaction is complete. These data also show that the increase in critical temperature depends on the initial microstructure. The fine quenched-and-tempered, or martensitic, microstructure revealed the least change in Ac_3 temperature compared with the equilibrium A_3 temperature, whereas the same steel with an annealed microstructure exhibits the largest difference in Ac_3 compared with the A_3 value obtained from very low heating rates. Such a trend is readily explained by the fact that the diffusion distance required for redistribution of carbon is shorter in the former instance and longer in the latter microstructure in which the carbides are much larger.

Data similar to those for 1042 steel are depicted in Fig. 5.2 and 5.3. For eutectoid 1080 carbon steel, the Ac_1 and Ac_3 temperatures nearly coincide in "slow" heat treatments ($Ac_1 \approx Ac_3 \approx A_1 \approx A_3$). By contrast, they differ for

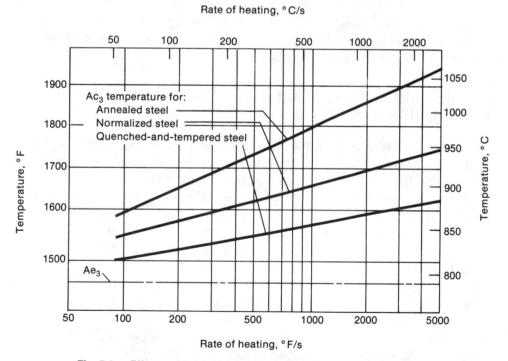

Fig. 5.1.   Effects of initial microstructure and heating rate on Ac_3 temperature for 1042 steel (Ref 55)

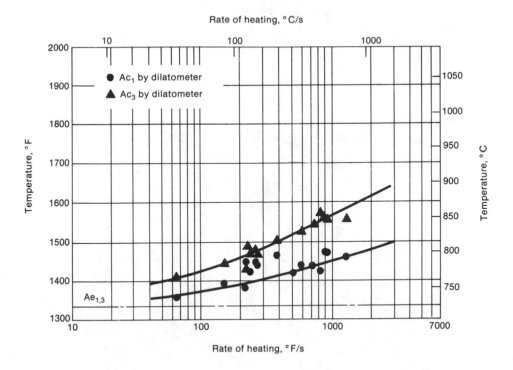

Fig. 5.2. Effect of heating rate on Ac₁ and Ac₃ temperatures for annealed 1080 steel (Ref 55)

rapid heating (Fig. 5.2) because of the way in which the experiments were conducted. The lower critical temperature can be taken as the one at which the reaction *started*, and the upper one as that at which transformation to austenite was *completed*. At the higher heating rates, Ac₁ and Ac₃ are higher than at lower rates, an effect which offsets the shorter times at temperature. The plots for the 4130 alloy steel (Fig. 5.3) reveal trends analogous to the previous ones for the carbon steel. The transformation temperatures for rapid heating are greater than those for slow heating, with the amount increasing as the structure becomes coarser. In addition, the difference between Ac₃ and Ac₁ for this steel is greater than that for the 1080 steel because of the passage through the $\alpha + \gamma$ phase field.

The above measurements are most useful when only the surface layers of a steel part are to be austenitized and hardened. In these cases, continuous rapid heating to the Ac₃ temperature is all that is needed. In other situations, in which deeper hardening or through-hardening is necessary, a certain amount of actual soaking time at temperature may be required. Examples of the time-temperature relationships have been presented previously in Chapter 2 (Fig. 2.6 and 2.7). Other examples are given in Fig. 5.4 for steel grades 1050, 1350, 2350, 4160, and 9255. The A_1 temperatures of these steels are approximately 725, 710, 690, 750, and 770 °C (1335, 1310, 1275, 1380, and 1420 °F), respectively. The corresponding

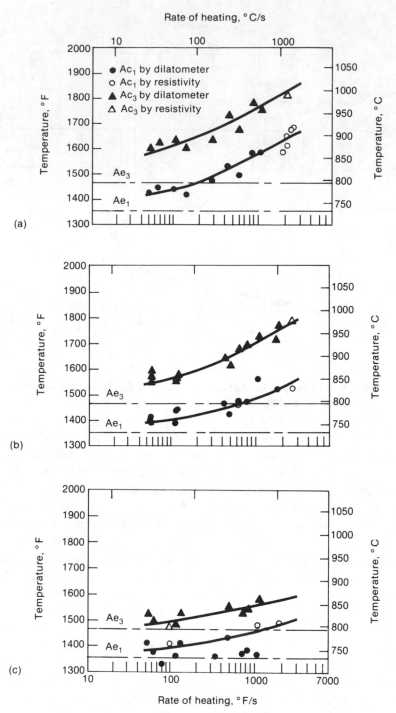

(a) Annealed. (b) Quenched and tempered at 675 °C (1250 °F).
(c) Quenched and tempered at 205 °C (400 °F).

Fig. 5.3. Effect of heating rate on Ac$_1$ and Ac$_3$ temperatures for 4130 steel (Ref 55)

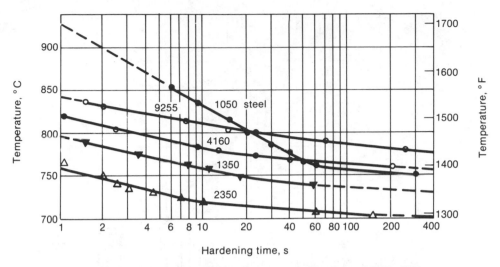

Fig. 5.4. TTT curves for complete austenitization of 1050, 1350, 2350, 4160, and 9255 steels each with a starting austenitized-and-furnace-cooled microstructure (Ref 56)

A_3 temperatures are 765, 750, 725, 775, and 800 °C (1410, 1380, 1335, 1425, and 1470 °F). Prior to austenitizing (in a lead bath), all steels had been previously austenitized and furnace cooled, resulting in a relatively coarse microstructure of pearlite and bainite with a large (1050) or small (1350 and 9255) amount of proeutectoid ferrite in the starting microstructure.

The data in Fig. 5.4 are for the times for complete reaction: for temperatures below the A_3 temperature, heat treatment involved partial austenitization, whereas above A_3 the microstructure was, of course, transformed entirely to austenite. Of importance is the fact that complete austenitization at or above the A_3 temperature required a finite amount of time. For example, for heat treatments carried out exactly at the A_3 temperature, reaction times of about 50, 18, 7, 22, and 22 s, respectively, are needed for steels 1050, 1350, 2350, 4160, and 9255 in the furnace-cooled condition. It is apparent that the steels with proeutectoid ferrite in the starting microstructure are the most difficult to austenitize. For that steel with the largest amount, 1050, the austenitizing time at A_3 can be shortened considerably with a starting bainitic microstructure (Fig. 5.5a). Such a modification reduces the required time from 50 to 8 s. Also of note in Fig. 5.4 and 5.5(a) is the fact that the austenitizing time can be substantially reduced by using temperatures somewhat higher than the A_3 temperature. For all steels except 1050, in the condition characterized by a large amount of ferrite in the starting microstructure, the time at temperature required is reduced to approximately 1 s for a temperature only 40 °C (70 °F) above the A_3 temperature. Even in the worst case—1050 steel containing a large amount of ferrite—it is possible to austenitize in about 1 s by raising the temperature about 160 °C (290 °F), or to a temperature still below that at which rapid grain growth occurs.

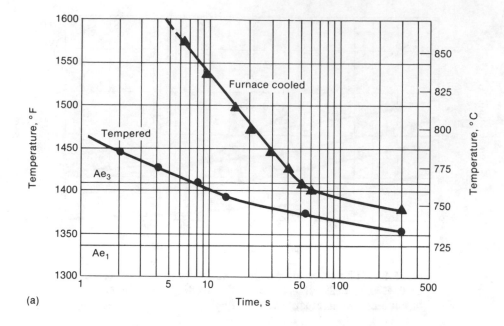

(a)

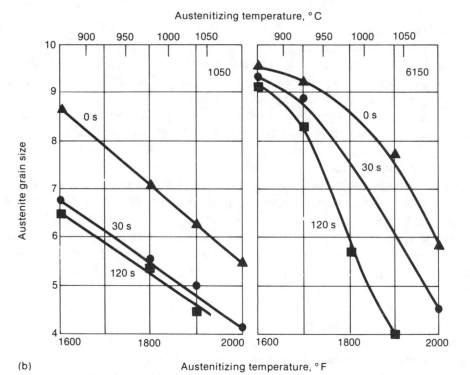

(b)

**Fig. 5.5. (a) Comparison of TTT curves for complete aus-
tenitization of 1050 steel with either an austenitized-and-furnace-
cooled or a quenched-and-tempered starting microstructure.
(b) Influence of austenitizing temperature and time on austenite
grain size in induction hardened 1050 and 6150 steels. (Ref 50, 56)**

The effect of time and temperature on austenite grain size is described quantitatively in Fig. 5.5(b). For times of 1 s, austenitizing temperatures must exceed 980 °C (1800 °F) in order for a relatively coarse grain size (ASTM 7) to be produced in 1050 steel. For another steel, 6150, with an A_3 temperature similar to that of 1050, temperatures in excess of 1050 °C (1925 °F) must be employed to yield a microstructure with equivalent grain size. The higher temperature is a result of the retarding effect of alloying elements (in this case, a 1% Cr addition, primarily).

The above trends for the austenitizing temperatures needed to minimize holding times, which are important from the viewpoint of rapid heating by induction, may be specified using a rule-of-thumb deduced from the above results. These temperatures for negligible holding times are summarized in Table 5.1 for plain carbon steels. Generally, the recommended temperatures increase with increasing A_3 or A_{cm} temperature; they are approximately 100 °C (180 °F) above the upper critical temperature primarily to keep the austenitizing time to a minimum. However, they are still below the temperature at which rapid growth of austenite grains occurs. In addition, recommended austenitizing temperatures are at least another 100 °C (180 °F) higher in alloys with strong carbide-forming elements (e.g., titanium, chromium, molybdenum, vanadium, or tungsten) than they are in carbon steels. These increases are the results of large increments in the critical temperatures of alloy steels. An insight into alloying effects and required induction hardening temperatures can be obtained from Fig. 2.4 and 2.5. At the same level of alloying, chromium gives rise to a moderate increase in the eutectoid and upper critical temperatures; tungsten, silicon, and molybdenum lead to larger increases; and titanium gives the largest increase. Two other common alloying elements, niobium and vanadium, can also be expected to raise the critical temperatures substantially; their effects are similar to that of titanium. Besides the increase in

Table 5.1. Recommended induction austenitizing temperatures for carbon and alloy steels (Ref 50)

| Carbon content, % | Induction austenitizing temperature(a) | | Quenchant |
	°C	°F	
0.30	900 to 925	1650 to 1700	Water
0.35	900	1650	Water
0.40	870 to 900	1600 to 1650	Water
0.45	870 to 900	1600 to 1650	Water
0.50	870	1600	Water
0.60	845 to 870	1550 to 1600	Water/oil
>0.60	815 to 845	1500 to 1550	Water/oil

(a) Free-machining and alloy grades are readily induction hardened. Alloy steels containing carbide-forming elements (e.g., niobium, titanium, vanadium, chromium, molybdenum, and tungsten) should be austenitized at temperatures at least 55 to 110 °C (100 to 200 °F) higher than those indicated.

recommended austenitizing temperatures in alloy steels, induction austenitizing times may also need to be slightly increased. This is because of the kinetics of alloy carbide dissolution in austenite, which can be substantially slower than those of cementite dissolution, particularly when NbC, TiC, and VC are involved. Such increases in induction hardening temperature and time generally are not deleterious from the viewpoint of austenite grain growth, however, because of the effects of alloying.

The above recommendations for induction hardening temperatures should be used only as a guide. This is especially true for modern plain carbon steels. These steels are typically melted using a large percentage of scrap which may contain microalloyed high-strength low-alloy steels. Thus, the steel may still be considered to be nominally of the plain carbon type but may contain trace amounts of niobium, vanadium, titanium, etc., whose presence may greatly affect heat treating response. For this reason, it is wise to obtain a complete chemistry on each lot of steel employed in induction hardening processes or to determine proper austenitizing temperatures experimentally.

One of the most interesting features of induction hardening is its ability to impart as-quenched hardnesses somewhat higher than those of conventionally furnace-hardened steels. This trait is illustrated in Fig. 5.6 for plain carbon steels

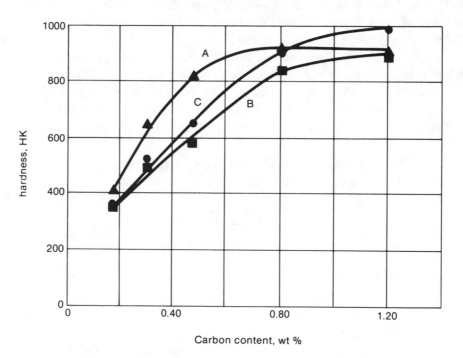

Curve A: induction hardened. Curve B: furnace hardened and water quenched. Curve C: furnace hardened, water quenched, and tempered. The quenched-and-tempered steels were treated in liquid nitrogen following water quenching prior to tempering at 100 °C (210 °F) for 2 h.

Fig. 5.6.   Effect of carbon content on hardness in plain carbon steels (Ref 57)

of various carbon contents. Data are presented for (A) surface induction hardened, (B) furnace through-hardened, and (C) furnace-hardened specimens given a low-temperature heat treatment consisting of cooling to liquid nitrogen temperatures and subsequently tempering at 100 °C (210 °F). The higher hardnesses of the induction hardened specimens may be attributable to three sources: residual stresses, smaller amounts of retained austenite, and carbon segregation. As to the first effect, compressive residual stresses are developed in surface-hardened steels because of the smaller density of martensite as compared with bainite or pearlite. During cooling following austenitizing, the higher-density inner layers shrink more than the surface layers, leading to such residual stresses (and thus increments in hardness). The second factor, smaller amounts of retained austenite in induction hardened steels, is a result of the finer martensite generally resulting from such steel heat treatments; the martensite may also be harder because it was formed from finer-grain, less-homogeneous austenite having a larger number of imperfections. The last factor, the increment in hardness due to carbon segregation, derives from the fact that induction austenitizing normally involves rapid heating and requires short holding times, which may lead to variation in carbon content within the austenite grains. Thus, a mixture of high-carbon and low-carbon martensite is formed during quenching. It is the high-carbon martensite which gives rise to higher hardnesses. This effect decreases in steels whose carbon contents exceed approximately 0.6 to 0.8 wt %, above which the hardness of martensite does not change (Fig. 2.26).*

ELECTRICAL AND MAGNETIC PROPERTIES OF STEEL

The principles underlying the induction heating of metals were outlined in Chapter 3. Here it was seen that the important material properties that determine the success of induction heating are the material's resistivity, ρ, and relative magnetic permeability, μ. Both of these factors enter the equation for reference depth, d (in.) $= 3160\sqrt{\rho/\mu f}$, where f is induction generator frequency in hertz (Hz) and ρ is resistivity in microhm-*inches* ($\mu\Omega \cdot$ in.). Since the values of ρ and μ are fixed once the material is selected, the only adjustable parameter is the frequency, f. Typically, higher frequencies are chosen for surface hardening and lower frequencies for through-hardening. More will be said about this shortly.

The resistivities of metals vary with temperature. Figure 5.7 shows this behavior for two ferrous alloys — electrolytic iron, an alloy with a negligible amount of carbon, and a 1% C steel. Both alloys have a similar dependence of resistivity on temperature. This can be ascribed to the fact that both consist largely of ferrite at low temperatures and austenite at high temperatures. In fact, a change in slope of

---

*It should be realized that rapid austenitizing by induction may produce substantial gradients in carbon content in the microstructure if time for homogenization is not allowed. The regions of very high carbon content may have M_f temperatures substantially below room temperature, thus leading to problems related to retained austenite. Tempering response of such a structure can be substantially different from the response of a structure consisting of a martensite of uniform carbon content which has been formed during a longer-time hardening operation.

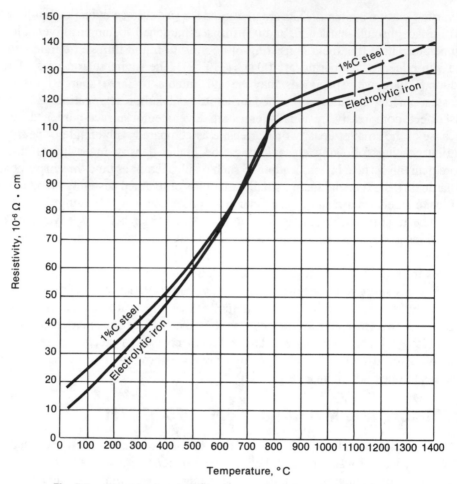

Fig. 5.7.   Electrical resistivity versus temperature for electrolytic iron and a 1%C steel (Ref 58)

the curves occurs at temperatures between 700 and 800 °C (1290 and 1470 °F) — a region in which phase transformations occur.

The temperature at which the slope of the resistivity-vs-temperature plot for steels changes also coincides with that at which magnetic properties show a related effect. It turns out that, as the temperature is increased, the relative magnetic permeability of steels decreases until a critical temperature known as the Curie temperature or Curie point is reached. Below this temperature, relative permeability varies with the intensity of the magnetic field and hence the current in carbon steels. It can attain values between 5 and 5000, approximately, at room temperature. Above the Curie point, however, permeability drops to unity, a value equal to the relative permeability of free space as well as of metals such as austenitic stainless steels, aluminum, and copper at *all* temperatures. The Curie temperature varies with composition in carbon steels. For steels with less than 0.45% C, this temperature is 770 °C (1420 °F). In higher-carbon steels, the Curie

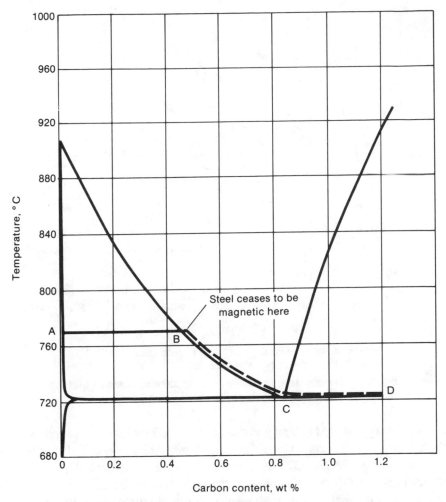

Fig. 5.8.   Curie temperature (line ABCD) of carbon steels (Ref 59)

temperature follows the A_3 line to the eutectoid composition; thereafter, it coincides with the A_1 line (Fig. 5.8). Alloying elements in steels change the Curie point by small amounts. Molybdenum and silicon increase it, and manganese and nickel decrease it. These effects are similar to those which these elements have on the eutectoid temperature (Fig. 2.5).

In physical terms, the decrease in magnetic permeability with temperature in carbon steels signifies the loss of ferromagnetic properties and some of the ease with which these steels may be heated by induction. Below the Curie point, heating occurs as a result of both eddy-current losses and hysteresis losses. The latter are no longer present once the ferromagnetic property of steels is eliminated. Such a consideration is very important with regard to the austenitizing of steels in the hardening operation since this is done above the Curie temperature. Figure 5.9 illustrates this effect. Here it will be noticed that, below the Curie point, the

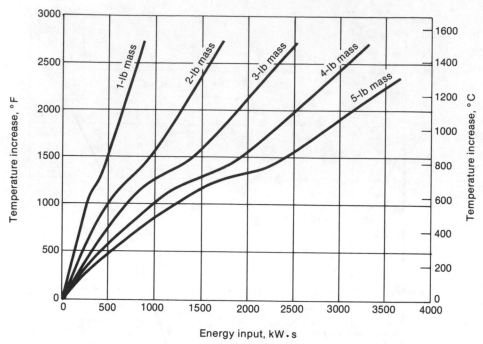

Note the decrease in heating rate as the Curie temperature (770 °C, or 1420 °F) is approached and exceeded.

Fig. 5.9.   Temperature increase as a function of energy input for induction heated carbon steel (Ref 51)

amount of energy needed to heat a given mass of steel by induction is proportional to the temperature increase. Once the Curie temperature is reached, the required amount of energy per degree of temperature increase is substantially larger. Once the steel is above this point, however, the necessary electrical energy, although reduced, is still greater than that needed for low-temperature heating. When austenitizing temperatures only slightly above the Curie temperature are to be used, no adjustments in induction equipment are usually necessary. When the temperatures must be relatively high, on the other hand, devices such as load-matching circuits may be employed. These vary the impedance of the power source to compensate for the change in the magnetic properties of the steel.

SELECTION OF FREQUENCY FOR HARDENING

The equation for reference depth, d, can be used to estimate the optimal generator frequency for induction hardening of steel. Assuming a relative magnetic permeability of unity in these cases and taking the resistivity to be $50 \times 10^{-6} \ \Omega \cdot \text{in.}$, which is an average value from Fig. 5.7, the expression for d becomes $d \, (\text{in.}) = 22.3\sqrt{1/f \, (\text{Hz})}$. For surface hardening, the desired case depth is typically taken to be equal to about one-half of the reference depth when selecting frequency. By contrast, when through-hardening is desired, the frequency is usually chosen such that the reference depth is a fraction of the bar radius (or an equivalent dimension for parts which are not round). This is necessary in order to maintain adequate "skin effect" and to enable induction to take place at all. If the reference depth is chosen to be comparable to or larger than the bar radius, there will be two "sets" of eddy currents near the center of the bar induced from diametrically opposed surfaces of the bar. These will tend to go in two different directions and, thus, cancel each other. To avoid this, frequencies for through-hardening are often chosen so that the reference depth does not exceed approximately one-fourth of the diameter for round parts or one-half the thickness for plates and slabs when using solenoid coils. When bar diameter is less than four reference depths, or slab thickness less than two reference depths, the electrical efficiency drops sharply, as has been discussed in Chapter 3. By contrast, little increase in efficiency is obtained when the bar diameter or slab thickness is many times more than the reference depth.

Typical frequency selections for induction hardening of steel parts are listed in Table 5.2 and Fig. 5.10. Those for surface hardening will be examined first. For very thin cases such as 0.38 to 1.27 mm (0.015 to 0.050 in.) on small-diameter bars, which are easily quenched to martensite, relatively high frequencies are optimal. If the reference depth is equated to the case depth, the best frequency for a 0.76-mm (0.030-in.) deep case on a 12.7-mm (0.5-in.) diameter bar is found to be around 550,000 Hz. When the surface of a larger-diameter bar is hardened, particularly when the case is to be deep, the frequency is often chosen so that the reference depth is several times the desired case depth. This is because the large amount of metal below the surface layer to be hardened represents a large thermal mass which draws heat from the surface. Unless very high power densities are employed, it is difficult to heat only the required depth totally to the austenitizing temperature. As an example, consider the recommended frequency for imparting a 3.81-mm (0.15-in.) hardened case to a bar 76.2 mm (3 in.) in diameter. If the reference depth were equated to the case depth, a frequency of about 20,000 Hz would be selected, which would provide only "fair" results. If a frequency of 3000 Hz were chosen, however, the reference depth would be about 10.3 mm (0.41 in.), or about 2½ times the required case depth. However, it is unlikely that the entire reference depth would ever reach austenitizing temperatures for the reason mentioned above.

For through-hardening of a steel bar or section, the optimal frequency is often based on producing a reference depth about one-fourth of the bar diameter or section size. For instance, through-heating and through-hardening of a 63.5-mm (2.5-in.) diameter bar would entail using a generator with a frequency of about 1000 Hz. If much lower frequencies were employed, inadequate skin effect (current "cancellation") and lower efficiency would result. On the other hand, higher frequencies might be used. In these cases, however, the generator power output would have to be low enough to allow conduction of heat from the outer regions of the steel part to the inner ones. Otherwise, the surface may be overheated, leading to possible austenite grain growth or even melting.

Table 5.2. Selection of power source and frequency for various applications of induction hardening (Ref 50)

Depth of hardening		Section size		Rating(a) for:					
				Power lines, 50 or 60 Hz	Frequency converter, 180 Hz	Solid-state systems or motor-generators			Vacuum tube, over 200 kHz
cm	in.	cm	in.			1000 Hz	3000 Hz	10,000 Hz	
Surface hardening									
0.038 to 0.127	0.015 to 0.050	0.64 to 2.54	1/4 to 1	...	...	...	...	...	Good
0.129 to 0.254	0.051 to 0.100	1.11 to 1.59	7/16 to 5/8	...	...	...	...	Fair	Good
		1.59 to 2.54	5/8 to 1	...	...	...	Fair	Good	Good
		2.54 to 5.08	1 to 2	...	...	Fair	Good	Good	Fair
		Over 5.08	Over 2	...	...	Good	Good	Good	Poor
0.257 to 0.508	0.101 to 0.200	1.91 to 5.08	3/4 to 2	...	...	Good	Good	Fair	Poor
		5.08 to 10.16	2 to 4	...	...	Good	Fair	Poor	...
		Over 10.16	Over 4	...	...	Good	Fair	...	...
Through-hardening									
		0.16 to 0.64	1/16 to 1/4	...	...	...	...	...	Good
		0.64 to 1.27	1/4 to 1/2	...	...	...	...	Fair	Good
		1.27 to 2.54	1/2 to 1	...	...	Fair	Fair	Good	Fair
		2.54 to 5.08	1 to 2	...	...	Good	Good	Fair	...
		5.08 to 7.62	2 to 3	...	...	Good	Good	Poor	...
		7.62 to 15.24	3 to 6	Fair	Good	Poor	Poor	Poor	...
		Over 15.24	Over 6	Good	Fair	Poor	Poor	Poor	...

(a) "Good" indicates a frequency that will most efficiently heat the material to austenitizing temperature. "Fair" indicates a frequency that is lower than optimum but high enough to heat the material to austenitizing temperature relative to the section size. With such a frequency, the current penetration relative to the section size causes current cancellation and lowered efficiency. For through-hardening, "Fair" may indicate a frequency higher than optimum that can result in overheating of the surface at high energy inputs. Converters cost more per kilowatt-hour than the converters of optimum frequency. With some equipment, the efficiency may be lower. "Poor" indicates a frequency that will result in overheating of the surface unless a low energy input is used. Efficiency and production are low, and capital cost of converters per kilowatt-hour is high.

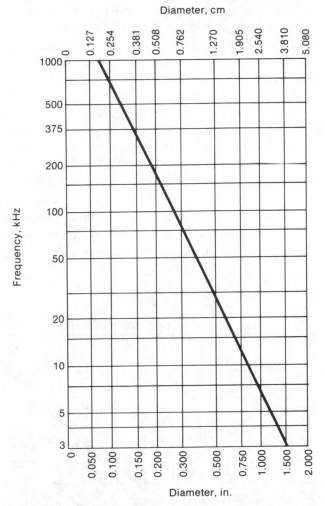

Fig. 5.10. Relationship between diameter of round steel bars and minimum generator frequency for efficient austenitizing using induction heating (Ref 51)

SELECTION OF POWER FOR HARDENING

The second major parameter in the selection of induction equipment for hardening of steel concerns power requirements. When only surface hardening is to be done, power density is usually high and heating times very short. In contrast, induction generators for through-hardening usually provide much lower power densities. This is because surface heating rates must be kept low enough to enable heat to be conducted to the center of the workpiece.

Induction generators are rated in terms of power, or the *rate* at which energy can be supplied to the coil (and subsequently to the workpiece). Therefore, the initial choice of power rating (in kilowatts) is based on the amount of heat energy (in kilowatt-hours) needed to heat the part to the required temperature and on the time available (in hours or, in the case of rapid heating, seconds). The amount of energy is easily estimated by realizing that the I^2R electrical energy dissipated by eddy currents in the workpiece is just another form of energy. Thermal energy is usually quoted in terms of Btu's or calories. One Btu (British thermal unit) is equivalent to 0.000293 kW · h; one calorie is equal to 1.163×10^{-6} kW · h.

The thermal energy, Q, needed to raise the temperature a certain amount, ΔT, depends on the weight or mass to be heated, m, and on the specific heat, c. This dependence is given by $Q = mc\Delta T$, with Q in Btu's (or calories), m in pounds (or grams), c in Btu/lb · °F (or cal/g · °C), and ΔT in °F (or °C). Often, the heat required per unit mass is dealt with. In this case, the quantity mc in the relation is replaced by c_H, or the *heat content* of the material. This is often a more useful quantity to use because the specific heat c varies with temperature. By using c_H, the amount of heat energy *per pound* of material required to raise the temperature to a given value may be found from plots such as that shown in Fig. 5.11. For

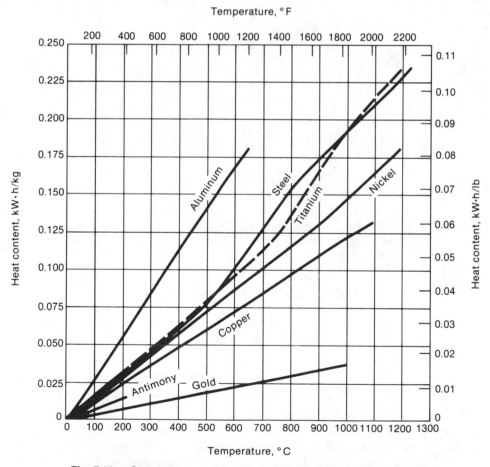

Fig. 5.11. Approximate heat content above 20 °C (70 °F) for various metals (Ref 43)

steel, for example, approximately 0.075 kW · h (255 Btu) are needed to increase the temperature of 1 lb of material from room temperature to an austenitizing temperature of 871 °C (1600 °F).

For surface hardening of steel, the power rating is chosen so that the required energy is supplied to the surface layers in several seconds or less. As might be expected, longer-duration heating is used when thicker cases are to be produced. Following heating, the quench is applied at a time interval determined by (a) time-temperature metallurgical considerations, such as those discussed earlier in this chapter, and (b) the rate at which heat is lost to the interior of the part to be surface hardened. Exact dwell times prior to quenching are very difficult to determine theoretically. Because of this, the best heating-and-quenching sequence is often found by trial-and-error.

Typical power ratings for surface hardening of steel are given in Table 5.3. These are based on the need to heat to temperature very rapidly and have proven to be appropriate through years of experience. Two examples will serve to show that these numbers are reasonable. First, consider the induction hardening of the surface of a small-diameter shaft using a 500-kHz generator. For a case depth of 0.76 mm (0.03 in.), heating of 1 in.^{2} of surface to 871 °C (1600 °F) requires 0.00064 kW · h (2.18 Btu), or 2.29 kW · s, of energy. Thus, a generator which can supply 1.55 kW/cm^2 (10 kW/in.^{2}), operating at 100% efficiency, can heat the surface of the shaft in 0.23 s. In actuality, it will take several times as long because of generator and coil losses as well as the fact that some of the power is used to heat the inner layers of the workpiece to temperatures below the transformation

Table 5.3. Power densities required for surface hardening of steel (Ref 50)

Frequency, kHz	Depth of hardening(a)		Input(b)(c), kW/in.^{2}(d)		
	cm	in.	Low(e)	Optimum(f)	High(g)
500	0.038 to 0.114	0.015 to 0.045	7	10	12
	0.114 to 0.229	0.045 to 0.090	3	5	8
10	0.152 to 0.229	0.060 to 0.090	8	10	16
	0.229 to 0.305	0.090 to 0.120	5	10	15
	0.305 to 0.406	0.120 to 0.160	5	10	14
3	0.229 to 0.305	0.090 to 0.120	10	15	17
	0.305 to 0.406	0.120 to 0.160	5	14	16
	0.406 to 0.508	0.160 to 0.200	5	10	14
1	0.508 to 0.711	0.200 to 0.280	5	10	12
	0.711 to 0.914	0.280 to 0.350	5	10	12

(a) For greater depths of hardening, lower kilowatt inputs are used. (b) These values are based on use of proper frequency and normal over-all operating efficiency of equipment. These values may be used for both static and progressive methods of heating; however, for some applications, higher inputs can be used for progressive hardening. (c) Kilowattage is read as maximum during heat cycle. (d) 1 kW/in.^{2} = 0.155 kW/cm^2. (e) Low kilowatt input may be used when generator capacity is limited. These kilowatt values may be used to calculate largest part hardened (single-shot method) with a given generator. (f) For best metallurgical results. (g) For higher production when generator capacity is available.

temperatures. As another illustration, consider the case hardening of a larger-diameter shaft to a depth of 7.6 mm (0.3 in.) using a 1000-Hz generator. The energy required is ten times that for the previous example, or 3.55 kW · s/cm² (22.9 kW · s/in.²). Similarly, a power rating on the order of 15.5 kW/cm² (100 kW/in.²) would be required to heat the shaft in an identical time. However, since cost *and* the likelihood for surface overheating increase with power rating, a power of only 1.55 kW/cm² (10 kW/in.²) or slightly more would also be optimal in this instance. The time to heat to temperature at this power level would still be only 2.3 s, however, under idealized conditions, or probably around two to three times this period in practice.

For other case depths and frequencies, it may be noted from Table 5.3 that a power rating of about 1.55 kW/cm² (10 kW/in.²) of surface to be treated is the typical recommendation. When using this or another fixed power rating, it should not be forgotten, nevertheless, that longer heating times are required for deeper case depths. This fact is illustrated in Fig. 5.12, in which the experimentally

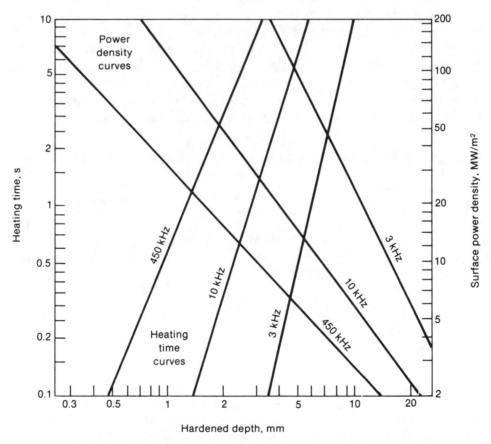

Fig. 5.12.   Interrelationship among heating time, surface power density, and hardened depth for various induction generator frequencies (Ref 60)

determined interrelation among power density, heating time, generator frequency, and hardened case depth is depicted. These graphs are based on typical efficiencies for induction setups and on heat losses due to conduction (beneath the hardened zone) and radiation. Moreover, the data are useful for installations in which parts are treated individually ("single-shot" techniques) as well as those in which heat treatment is done incrementally on different sections of a large component ("scanning" techniques, which will be discussed shortly). It should be noted, nevertheless, that equivalent case-hardened depths can be obtained using process parameters different from those in Fig. 5.12. In general, longer times at lower power densities will yield an identical case depth for a given frequency (Ref 61); however, in such situations, more heat is conducted into the core of the part, resulting in lower over-all system efficiency than is realized at higher power densities.

Power ratings for through-heating for hardening are much lower than those for surface hardening. As has been mentioned above, this allows relatively slow heating of the surface layers and time for the heat to be conducted to the center of the workpiece. The variation of the surface and center temperatures of an induction heated workpiece is depicted schematically in Fig. 5.13. At the beginning of heating, the temperature at the surface increases much more rapidly than that at the center. After a while, the rates of increase of the surface and center temperatures become comparable due to conduction. However, a fixed temperature differential persists during heating. Using methods described by Tudbury (Ref 43), the allowable temperature differential permits the generator power rating to be selected. The basic steps in selecting the power rating are as follows:

- Select the frequency and calculate the ratio of bar diameter (or section size) to reference depth, a/d. For most through-heating applications, this ratio will be around four to six.
- Using the values of the thermal conductivity (in W/in. · °F) and a/d, estimate the induction thermal factor, K_T (Fig. 5.14).
- The power per unit length is calculated as the product of K_T and the allowable temperature differential (in °F) between the surface and center, $T_s - T_c$. Multiplying this by the length of the bar yields the net power required in kilowatts.

To illustrate the above procedure, consider the through-heating of a steel bar 152.4 mm (6 in.) long and 101.6 mm (4 in.) in diameter to a surface temperature of 980 °C (1800 °F) and a center temperature of 870 °C (1600 °F). For a 540-Hz generator, the a/d ratio is found to be approximately four. The average value of the thermal conductivity for carbon steels when heated to hardening temperature is about 0.7 W/in. · °F. From Fig. 5.14, K_T is found to be equal to 0.016. Thus, the power required is equal to $(0.016)(200)(6) = 19.2$ kW. This corresponds to a power density (power per unit of surface area) of $19.2/(15.2)(10.2)\pi = 0.039$ kW/cm^2 (0.25 kW/in.^{2}). If the temperature differential were to be cut by a factor of two, the power rating and power density would be similarly decreased. There is a lower limit to which the power can be decreased, though. This is

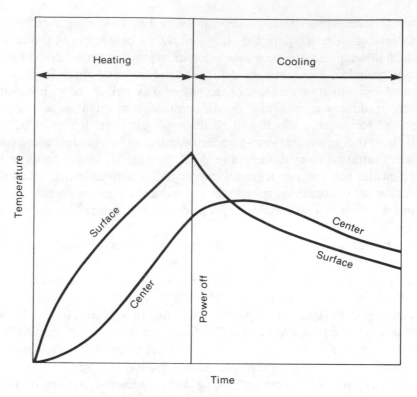

Note that, following an initial transient, the surface-to-center temperature differential is constant during the heating cycle. After
heating, however, the surface cools more rapidly, leading to a
temperature crossover.

**Fig. 5.13. Schematic illustration of the surface and center tem-
perature histories of a bar heated by induction (Ref 43)**

determined by the heating time itself. The heating time is equal to the amount of
energy needed divided by the rate of energy input, or power. For the above
example, the workpiece weighs approximately 9.66 kg (21.3 lb), which, using
data from Fig. 5.11, requires $(21.3)(0.082) = 1.75$ kW · h for heating to 925 °C
(1700 °F). Therefore, the heating time will be either 328 s (for a surface-to-center
temperature differential of 110 °C, or 200 °F) or 656 s (for a temperature differential of 55 °C, or 100 °F) when the power rating is halved.

It should be realized that all of the above calculations neglect heat losses arising
from radiation and electrical losses. Assuming black-body radiation, the former
losses can also be established using data from Tudbury's book (Fig. 5.15). This
type of radiation represents an upper limit on the losses by these means. In the
above example, the rate of energy loss for a surface temperature of 980 °C
(1800 °F) is 0.0133 kW/cm^2 (0.086 kW/in.^{2}). This should be added to the
0.0388-kW/cm^2 (0.25-kW/in.^{2}) power density required with no radiation, to
yield a value of approximately 0.052 kW/cm^2 (0.34 kW/in.^{2}). To this must be

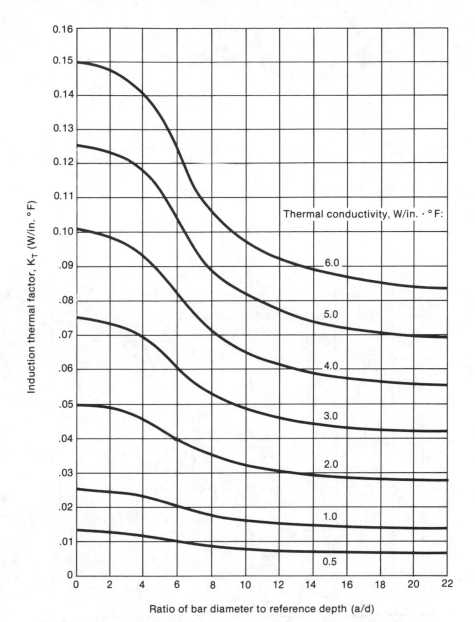

Fig. 5.14. **Induction thermal factor for round bars as a function of the ratio of bar diameter to reference depth (a/d) and the thermal conductivity (Ref 43)**

added the power to compensate for electrical losses. For a/d's of four to six, the efficiency of an induction system is usually around 75%. Therefore, the total power required is $0.052/0.75 \approx 0.069$ kW/cm^2 (0.45 kW/in.^{2}).

In order to avoid calculations such as the above, tables of power densities ordinarily used for through-heating of steel (for hardening as well as other uses,

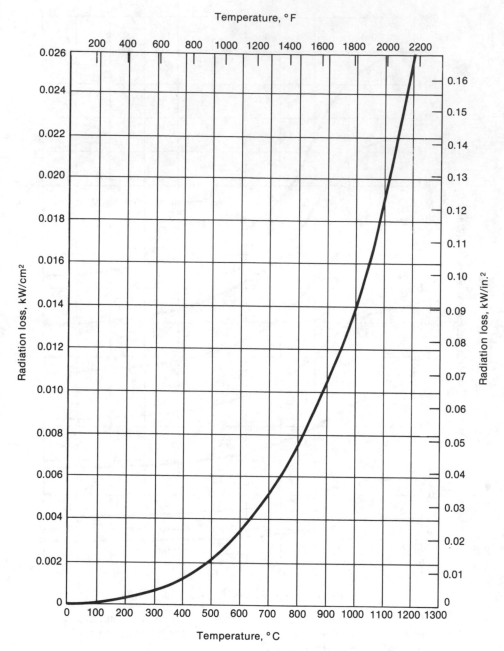

Losses are based on black-body radiation into surroundings at
20 °C (70 °F).

Fig. 5.15.   Radiation heat loss as a function of surface temperature (Ref 43)

such as forging) are available. One such listing is shown in Table 5.4. These
values of power densities are based on typical electrical efficiencies and proper
selection of frequency (which lead to a/d's usually in the range of four to six). It
may be noted that the larger-diameter bars, which can be heated efficiently with

Table 5.4.   Approximate power densities required for through-heating of steel for hardening, tempering, or forming operations (Ref 50)

Frequency(a), Hz	Input(b), kW/in.²(c)				
	150 to 425 °C (300 to 800 °F)	425 to 760 °C (800 to 1400 °F)	760 to 980 °C (1400 to 1800 °F)	980 to 1095 °C (1800 to 2000 °F)	1095 to 1205 °C (2000 to 2200 °F)
60	0.06	0.15	(d)	(d)	(d)
180	0.05	0.14	(d)	(d)	(d)
1,000	0.04	0.12	0.5	1.0	1.4
3,000	0.03	0.10	0.4	0.55	0.7
10,000	0.02	0.08	0.3	0.45	0.55

(a) The values in this table are based on use of proper frequency and normal over-all operating efficiency of equipment. (b) In general, these power densities are for section sizes of 1.27 to 5.08 cm (1/2 to 2 in.). Higher inputs can be used for smaller section sizes, and lower inputs may be required for larger section sizes. (c) 1 kW/in.² = 0.155 kW/cm². (d) Not recommended for these temperatures.

lower-cost, lower-frequency power supplies, typically employ smaller power densities than small-diameter bars. This is because of the greater times required for heat to be conducted to the center of the larger pieces. Also, it can be seen that lower frequencies such as 60 and 180 Hz are not ordinarily recommended for through-heating of steel when temperatures above approximately 760 °C (1400 °F) are desired. This is due to the increased reference depth (and decreased skin effect) above the Curie temperature where the relative magnetic permeability drops to unity. An exception to this practice is the use of 60-Hz sources for induction heating of very large parts such as steel slabs in steel mills.

One mitigating effect which must be considered when establishing power requirements for austenitizing is the delay between the time at which the power is turned off and the time at which the quench is applied. As shown in Fig. 5.13, following heating, the temperature at the surface drops more rapidly than that at the center of the workpiece. Eventually, the center temperature becomes greater. Because of this, the heating and cooling cycles can often be adjusted to compensate for the nonuniform heating which characterizes induction processes. Thus, greater input power and higher heating rates can sometimes be realized than when quenching follows immediately after heating.

SELECTION OF HEATING METHOD: SINGLE-SHOT VERSUS SCANNING TECHNIQUES FOR THROUGH-HEATING

When only one part at a time is placed in the coil and austenitized, the process is referred to as a "single-shot" process. Although handling of parts and conversion of the system for treatment of parts of different sizes are easy with this arrangement, such systems frequently are not economical to operate because the induction generator is not operated continuously at its rated power output. This type of system is not commonly used for long-time heating for through-hardening.

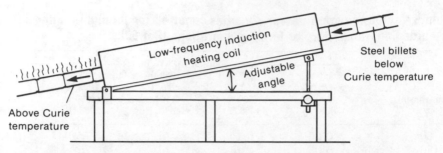

Fig. 5.16.   Forces (arrows) on induction heated (magnetic) steel bars and typical setup for continuous heating prior to quench-hardening or metalworking (Ref 43)

Another class of technique which allows high output in production environments, particularly for through-heating applications, is that referred to as "induction scanning." In this method, the workpiece is moved through the coil. Alternatively, a short *coil* may be moved along the workpiece length once (or several times), a common procedure when the workpiece is large and not easily moved. In this manner, the generator is operated continuously, and, if its power output is low, smaller areas can still be heat treated, leading to progressive processing of the part.

Simultaneous heating of multiple billets is one application classified under the broad heading of the induction scanning methods. In these instances, heating is done continuously, following which the workpiece may be quenched or sent to another work station such as a forging press. A typical setup is illustrated schematically in Fig. 5.16. Here, steel bars are loaded onto a conveyor which takes them down into the long coil. Ten or more billets may be in the coil at any one time, ensuring that one comes out every few seconds. The exact speed at which they do exit is, of course, determined by the power input and heating rate. However, once the necessary transit time is estimated, it can be controlled by adjusting the angle which the coil forms with the ground. Such an adjustment allows for balancing of gravitational and magnetic forces such that the steel billets essentially move by themselves. The gravitational force, of course, is always acting downwards. On the other hand, the magnetic force changes once the Curie temperature is reached. Below this temperature, where the steel is magnetic, billets are drawn into the coil. Once the billet temperature exceeds the Curie temperature and the billet becomes nonmagnetic, the billet is repelled by the coil. This repulsive force, in conjunction with the "adjustable" gravitational force and the small vibrations induced by the heating method itself, are enough to cause the bar to be pushed from the coil.

The use of long heating trains, such as is often done for heating of multiple steel billets, may actually involve more than one coil. These coils may be attached to several generators of different frequencies. For instance, the coil near the entrance to the heating line may be powered by a relatively low-frequency source. As the Curie point is approached and exceeded, however, steel billets may be passing through a coil connected to a somewhat higher-frequency source, a procedure which maintains sufficient skin effect for efficient operation.

Typical operating conditions for through-thickness induction austenitizing using scanning techniques are given in Table 5.5. As before, the power-density ratings are rather low (usually in the range of 0.5 to 2 kW/in.^{2}); the frequent use of dual frequency may be noted. The data in the table show that the changeover from low to high frequencies may also occur at temperatures below the Curie point, a practice used to compensate for the increase in resistivity (or gradual decrease in magnetic permeability) with temperature that steel exhibits (Fig. 5.7). In addition, relatively long heating times are still obvious even with induction scanning. As in most induction heat treating operations, however, these times are still an order of magnitude (or more) shorter than those for furnace treatments for which 30 min to 1 h per inch of thickness is often used for austenitizing.

QUENCHING TECHNIQUES AND QUENCHANTS FOR INDUCTION AUSTENITIZED PARTS

Quenching techniques are as important a design feature of induction hardening lines as the equipment and coil used for austenitizing. The important questions to be answered when determining quenching systems include the following:

- Part size and geometry
- Type of austenitizing operation (surface or through-hardening)
- Type of heating method (single-shot or scanning)
- Hardenability of steel and quenchant needed.

The two most common types of systems consist of spray quench rings and immersion techniques. When quench rings are used for round bars, their shape, like the coil, is generally round. The ring may be located concentric with the coil (Fig. 5.17) or directly underneath or alongside it (Fig. 5.18a) as in single-shot induction hardening setups. In those using induction scanning, parts move through the quench ring and coil, with quenching occurring immediately after heating (Fig. 5.18b). For nonsymmetric parts, the quenching apparatus, like the coil, is generally of the same shape as the part. As an example, Fig. 5.19 shows a specialized fixture for quenching the induction austenitized teeth on the internal diameters of gears.

In addition to the coil-and-quench-ring arrangements mentioned above, several other systems are in common use. These have been summarized by Spencer and his colleagues (Ref 50) and are depicted in Fig. 5.20. In brief, they consist of the following systems and operations:

- Heat in coil; remove part manually and quench.
- Heat and quench in one position (single-shot method described above).
- Heat in coil with part stationary; quench ring moves into place (modified single-shot/scanning technique).
- Heat in coil; part lowered automatically into agitated quench tank.

Table 5.5. Typical operating conditions for progressive through-hardening of steel parts by induction (Ref 50)

| Section size | | Material | Frequency(a), Hz | Power(b), kW | Total heating time, s | Scan time | | Work temperature | | | | Production rate | | Inductor input(c) | |
cm	in.					s/cm	s/in.	Entering coil °C	°F	Leaving coil °C	°F	kg/h	lb/h	kW/cm²	kW/in.²
Rounds															
1.27	1/2	4130	180	20	38	0.39	1	75	165	510	950	92	202	0.067	0.43
			9600	21	17	0.39	1	510	950	925	1700	92	202	0.122	0.79
1.91	3/4	1035 mod	180	28.5	68.4	0.71	1.8	75	165	620	1150	113	250	0.062	0.40
			9600	20.6	28.8	0.71	1.8	620	1150	955	1750	113	250	0.085	0.55
2.54	1	1041	180	33	98.8	1.02	2.6	70	160	620	1150	141	311	0.054	0.35
			9600	19.5	44.2	1.02	2.6	620	1150	955	1750	141	311	0.057	0.37
2.86	1⅛	1041	180	36	114	1.18	3.0	75	165	620	1150	153	338	0.053	0.34
			9600	19.1	51	1.18	3.0	620	1150	955	1750	153	338	0.050	0.32
4.92	1¹⁵⁄₁₆	14B35H	180	35	260	2.76	7.0	75	165	635	1175	195	429	0.029	0.19
			9600	32	119	2.76	7.0	635	1175	955	1750	195	429	0.048	0.31
Flats															
1.59	5/8	1038	3000	300	11.3	0.59	1.5	20	70	870	1600	1449	3194	0.361	2.33
1.91	3/4	1038	3000	332	15	0.79	2.0	20	70	870	1600	1576	3474	0.319	2.06
2.22	7/8	1043	3000	336	28.5	1.50	3.8	20	70	870	1600	1609	3548	0.206	1.33
2.54	1	1036	3000	304	26.3	1.38	3.5	20	70	870	1600	1595	3517	0.225	1.45
2.86	1⅛	1036	3000	344	36.0	1.89	4.8	20	70	870	1600	1678	3701	0.208	1.34
Irregular shapes															
1.75 to 3.33	¹¹⁄₁₆ to 1⁵⁄₁₆	1037 mod	3000	580	254	0.94	2.4	20	70	885	1625	2211	4875	0.040	0.26

(a) Note use of dual frequencies for round sections. (b) Power transmitted by the inductor at the operating frequency indicated. This power is approximately 25% less than the power input to the machine, because of losses within the machine. (c) At the operating frequency of the inductor.

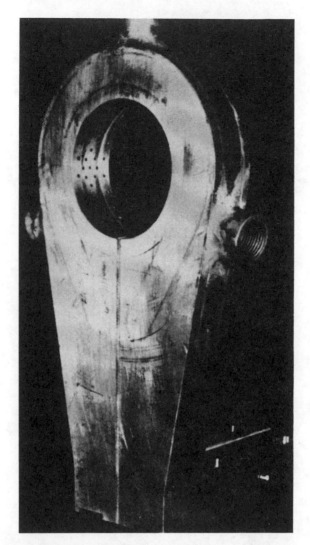

Fig. 5.17.   Typical quench ring for hardening of induction austenitized round parts (Ref 62)

- Heat part in vertical or horizontal position using scanning technique with *integral* quench ring (described above).
- Heat part using scanning technique and separate multirow quench ring.
- Heat in coil; part self-quenched (i.e., surface-heated layers quenched by cooler internal layers) or quenched with compressed air. Used primarily for high-hardenability steels.
- Parts pushed through coil and dropped onto submerged quench conveyor.
- Part with flanged end heated by scanning technique; shaft quenched continuously during scanning and flange quenched by second quench orifice at end of scan.

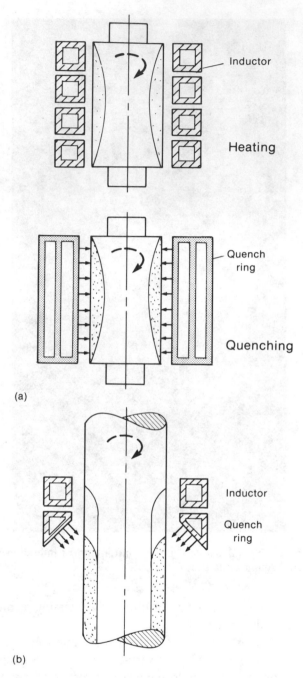

Note that round parts are usually rotated during heat treatment in order to enhance temperature uniformity.

Fig. 5.18. Typical setups for austenitizing and quenching in (a) single-shot and (b) scanning induction hardening processes (Ref 63)

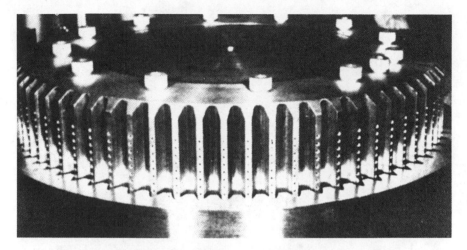

Fig. 5.19.   Device designed to fit inside a gear in order to quench inner-diameter gear teeth during induction hardening (Ref 62)

- Part heated vertically by scanning technique and dual quenched by integral quench ring and submerged spray quench.
- Part (such as crankshaft bearing surface) heated by split inductor and integral split quench ring.

Sufficient quenchant flow must be maintained to cool the part or section being quenched. Because induction heating systems are themselves compact, quenching systems are frequently designed smaller than they should be. To avoid this, the capacity of the pumping system should be at least three or four times the flow rate needed for proper quenching, and the quenching flow rate should be adjusted so that quenchant does not boil off once the part leaves the quench-ring location. Furthermore, if part rotation is used during heating and quenching, its rate must be kept low enough to avoid excessive quenchant from being thrown off.

As with furnace heat treatments, water and oil are frequently used as quench media in induction heat treatment practice. Water is the more common. Oil is typically used only when heat treatment is to be performed on steels of high hardenability, or on parts in which cracking or distortion is likely to occur.

When water is used, it is best to select a supply which is reasonably clean and not extremely hard. Dirt may tend to clog the orifices of the quench tooling; similarly, hard-water deposits, which may build up slowly in quench rings, cut down on their efficiency and may necessitate replacement or extensive cleaning. Besides cleanness requirements, the water temperature should also be controlled, preferably in the range of 15 to 40 °C (60 to 105 °F). This is most easily done when the water supply is large and when specialized recirculating systems are used.

Oils for quenching come in three generalized categories: general-purpose quenching oils (paraffin-type oils), "fast-quenching" oils, and soluble oil-water mixtures. Care must be exercised with all of these oils to provide adequate ventilation for removal of oil vapors from the air and, thus, to prevent flash fires.

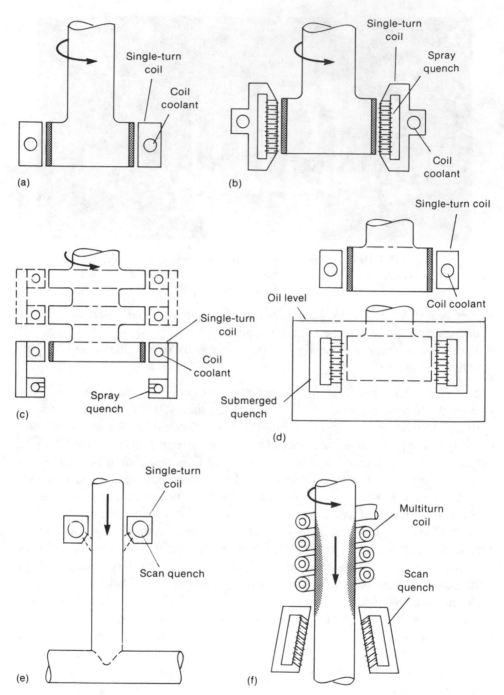

Fig. 5.20.   Common arrangements for quenching of induction heated steel parts (Ref 50)

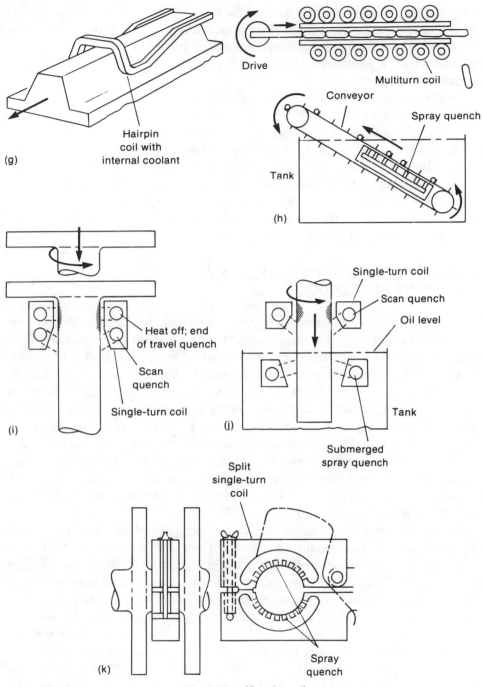

Fig. 5.20. (Continued)

This is especially important with some of the low-flash-point, fast-quenching oils. The best way to control and minimize the presence of oil vapors is to supply a large amount of oil which completely covers the heated portion immediately below the inductor until the temperature of the quenched area is below the vaporization temperature of the oil.

Equally common and successful quenchants for induction hardening applications include polyvinyl alcohol solutions and compressed air. The former have become very popular in recent years and are used in hardening parts with borderline hardenability for which oil does not cool fast enough and water quenching leads to distortion or cracking. Unlike oil, polyvinyl alcohol, one of the so-called polymer quenchants, is not flammable and does not produce objectionable fumes or irritate the skin. Compressed-air quenching is typically used for high-hardenability, surface-hardened steels from which relatively little heat needs to be removed. Typical applications include gear teeth.

QUENCH CONTROL IN INDUCTION HEAT TREATMENT OPERATIONS

Adequate controls are necessary to ensure consistent results in induction heat treatment. Such controls for spray quenches include those for quenchant flow, temperature, timing, etc. The over-all flow rate, *per se*, is controlled by adjustments of the pumping system itself. However, other considerations such as quench-device coupling and hole spacing in the quench device are also important (Ref 62 and 64). Often in single-shot operations, coupling — or the distance between the quench ring and part to be hardened — is very close, sometimes as little as 1.3 mm (0.051 in.). When several different sizes are to be treated, however, this distance may vary. Large distances are not desirable, however, because the velocity of the quenchant stream drops as the stream lengthens. This explains why the quench ring is often contoured to the part, i.e., to maintain uniform coupling between the two.

The size and spacing of the orifices in the spray-quench device are a second important consideration in quench control. These can be adjusted to produce uniform quenching and avoid cracking or soft spots. Usually, this involves designs with many small holes rather than a few large ones. The lower limit on size is the minimum size that can be drilled on a production basis, as well as that which can be kept free of dirt. Also, smaller pieces heat treated by single-shot methods generally require smaller holes than large ones. Typically, for parts 0.64 to 1.27 cm (0.25 to 0.50 in.) in diameter, orifice holes should be no larger than 0.16 cm (0.06 in.) in diameter. For parts 1.27 to 3.81 cm (0.5 to 1.5 in.) in diameter, orifice sizes of about 0.32 cm (0.125 in.) are often employed. The largest orifice diameter is usually 0.64 cm (0.25 in.). Exceptions to these rules apply to scanning heat treatment, in which the motion between the quench device and part, as well as rotation of the part, help distribute the flow. Thus, in these cases, larger holes can be employed.

Spacing of the holes in the quench device must be such that each one has approximately the same amount of area to quench. Sometimes, staggering or overlapping of these "spheres of influence" is useful to be sure that the entire part is quenched. In practice, however, spacing may vary although the ratio of quench-device surface area to orifice area is typically about 10 to 1 or 20 to 1 for setups with narrow or wide coils, respectively. In either case, however, additional holes are placed at the end in order to provide quenchant flow to the ends of parts in single-shot applications.

The number of rows of holes in the quench device depends on the cooling rate necessary to harden the steel, the depth of heating (surface or through), the properties of the quench medium, the rate of travel (in scanning operations), and the configuration of the part. The desired surface finish may even affect the quench design and quench action. One of the major differences between quench rings for single-shot and scanning arrangements lies in the angular orientation of the holes. Single-shot spray-quench devices have orifices which are perpendicular (i.e., radial) to the axis of the part. By contrast, the optimal angle between the axis of the holes and part axis in induction scanning setups is 30°. This angle is selected to allow sufficient quenching action while at the same time preventing the quenchant from interfering with the heating part of the operation.

It may also be necessary to install baffles in the quench chamber in order to ensure uniform flow along the entire surface. Baffles may take several forms, including simply a plate in front of the inlet hole (arranged so that the quenchant hits the plate rather than the quench-device orifices closest to the inlet hole) or a complete ring or plate between inlet and outlets with holes two to four times the diameter of the quench orifices. Among the considerations governing the need for baffles in the quench device are the following:

- Relative positions of the inlet hole and outlet orifices and whether the axes of both are parallel or perpendicular to each other.
- Quench pressure and inlet size. Lower inlet pressures and large inlet diameters necessitate the use of baffling.
- Orifice size. Baffling is more important as orifice size increases.

Control of the temperature of the quenching medium and its timing are also important. To this end, heat exchangers or cooling towers are often integral parts of an induction heating installation. In addition, electronic controls are often used to maintain timing of the quench cycle. This is particularly important when a tempering operation directly follows the hardening one or a so-called "auto-tempering" process is employed. In the former instance, the quench duration is typically adjusted so that the part temperature does not drop completely to room temperature. Leaving a small amount of residual heat in the component makes subsequent induction heating easier but does not affect the quality of the temper. In the latter case, the quench time may be controlled to bring the part temperature down precisely to tempering temperature, for instance, when a tempered pearlite or a tempered bainite microstructure is needed. Such an operation frequently involves precise temperature monitoring equipment and electronic feedback circuitry. One of the most interesting applications of this technique is the hardening and tempering of railroad rails (Ref 65). Following austenitizing, the surface of the rail is air quenched to 425 °C (800 °F). Subsequently, the surface temperature climbs back to 595 °C (1100 °F) because of residual heat from the interior of the rail, leading to autotempering. Finally, a sustained cold-water quench is applied to bring the entire rail down to room temperature.

When quenching is done improperly, several problems may arise, including soft spots, quench cracks, and part distortion. Soft spots sometimes occur when water is used as the quenchant; they result from the formation of steam pockets on the part surface which prevent rapid enough cooling for the formation of martensite. As might be expected, this problem is most severe in low-hardenability steels and can be alleviated by improved quench ring design or changes in the quenching device/part configuration. Quench cracks are typically due to one or more of four separate factors: (1) excessive quench severity (which is particularly troublesome in higher-carbon steels); (2) nonuniformity of quenching; (3) changes in part contours with insufficient transitional areas; and (4) surface roughness (e.g., tool marks). Part distortion is commonly caused by relief of residual stresses, uneven heating, nonuniform quenching, or part geometry. In many cases, these can be controlled by inductor changes, quench baffling, or modifications in part-handling techniques during heating and quenching.

HARDNESS DISTRIBUTIONS IN INDUCTION HARDENED PARTS

Several examples will serve to show that induction hardening can produce results as good as those of conventional furnace heat treatments. The first example (Fig. 5.21) shows the uniformity that can be obtained in through-hardening of steel bar stock. The data are for a carbon steel and two alloy steels each initially 31.8 mm (1.25 in.) in diameter. It can be seen that the hardness of the two alloy steels is relatively uniform, particularly in the 4140 steel. There is, however, a hardness gradient in the 1045 steel which is most likely attributable to a lack of hardenability rather than to poor or nonuniform austenitizing. In fact, it is surprising that the uniformity is as good as it is when this hardness profile is compared with that in a conventionally heated bar of the same material which was quenched in a *Jominy* fixture (Fig. 2.14).

The effect of varying the induction heating variables for the through-hardening of a high-hardenability steel, 4340, is shown in Fig. 5.22 and 5.23, with data obtained by Poynter (Ref 67). In all cases, bars 25.4 mm (1 in.) in diameter were induction heated using a 66.7-mm (2.63-in.) long coil and spark-gap generator (with a frequency of 355 kHz). The input power rating of the generator was 30 kW, resulting in a theoretical maximum power density of 0.56 kW/cm^2 (3.6 kW/in.^{2}). Figure 5.22 gives results of induction heating trials when the steel was in the normalized-and-tempered condition, or one in which austenitizing was

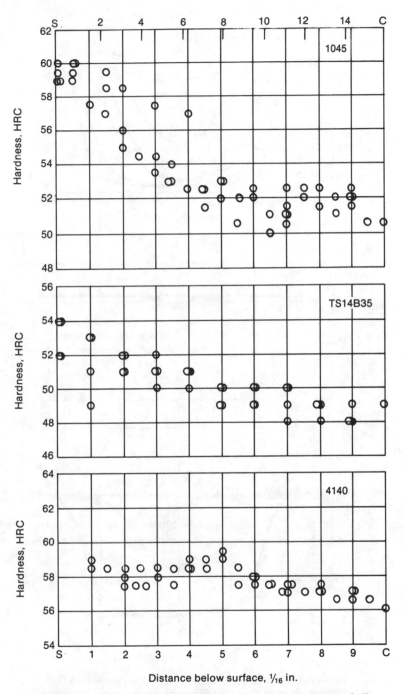

Fig. 5.21.   Typical hardness profiles across 3.18-cm (1.25-in.) diameter bars of various steels through-hardened by induction processes (Ref 50)

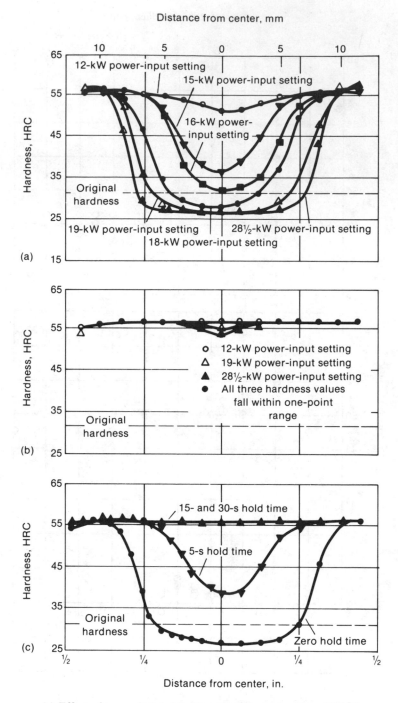

(a) Effect of power input. Maximum surface temperature, 815 °C (1500 °F); no hold time. (b) Effect of power input. Maximum surface temperature, 870 °C (1600 °F); no hold time. (c) Effect of hold time on reaching a surface temperature of 815 °C (1500 °F). Power input, 19 kW.

Fig. 5.22. **Effects of induction heating variables on as-quenched radial hardness distributions in 4340 steel bars 2.54 cm (1 in.) in diameter and 11.43 cm (4.5 in.) long with a normalized-and-tempered starting microstructure (Ref 66)**

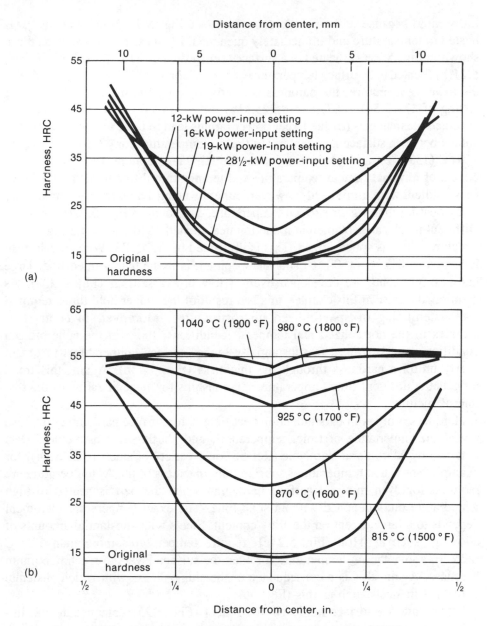

Distance from center, mm

(a)

(b)

Distance from center, in.

(a) Effect of power input. Maximum surface temperature, 815 °C (1500 °F); no hold time. (b) Effect of maximum surface temperature. Power input, 19 kW; no hold time.

Fig. 5.23.   Effects of induction heating variables on as-quenched radial hardness distributions in 4340 steel bars 2.54 cm (1 in.) in diameter and 11.43 cm (4.5 in.) long with a spheroidized starting microstructure (Ref 66)

fairly rapid because of a fine microstructure (see Fig. 5.1). When the steel was heated to temperature and immediately quenched, hardness profiles such as those shown in Fig. 5.22(a) (heated to a surface temperature of 816 °C, or 1500 °F) and 5.21(b) (heated to a surface temperature of 870 °C, or 1600 °F) were obtained. For the lower temperature, the hardness is relatively uniform only at the low power setting of $12/(2.54)(6.67)\pi = 0.225$ kW/cm^2 (1.45 kW/in.^{2}). This is not unexpected inasmuch as (a) the larger power ratings lead to large temperature differentials between surface and center; (b) the A_3 temperature for 4340 is only about 750 °C (1380 °F); and (c) the time at austenitizing temperature was minimal. The effects of minimal time at temperature and temperature differentials on the hardness gradient were minimized, however, either by using a hold time at temperature or by heating to a higher surface temperature, as shown in Fig. 5.22(b) and (c). Different results are obtained in induction heat treated 4340 with a starting microstructure which is spheroidized. This is illustrated in Fig. 5.23. When the surface is heated to 815 °C (1500 °F) and the specimen is immediately quenched, large hardness gradients are developed even at low power settings (Fig. 5.23a). As mentioned earlier in this chapter, this is a result of the longer hold times required for austenitizing of spheroidized microstructures. Moreover, in contrast to samples in the normalized and tempered conditions, those in the spheroidized condition must be raised to substantially higher surface temperatures to obtain nearly uniform hardness through the thickness (Fig. 5.23b). Again, this trend reflects the different time-temperature relationships for austenitization of the two microstructures.

Hardness profiles in bars induction heat treated for surface hardening only also reveal the importance of time, temperature, and starting microstructure. Data obtained by Martin and Van Note (Ref 56) on a variety of steels (Fig. 5.24), for example, show that temperatures substantially in excess of the A_3 temperature are required for even minimal surface hardening when the part is heated to high temperature and quenched without soak time. The exact temperature increment depends to a large extent on the alloy content. Steels with substantial amounts of alloying, such as 4160 (Fig. 5.24c), require temperatures more than 125 °C (225 °F) above the A_3 temperature for even the surface layers to reach maximum hardness. Again, this is a consequence of the sluggishness with which austenite is formed in steels such as this (Fig. 5.4).

Other data for surface-hardened 1070 steel (Fig. 5.25) demonstrate the important effect of starting microstructure on case depth for steels heated to temperature and quenched immediately. These results illustrate the fact that steels with quenched-and-tempered starting microstructures tend to develop much deeper hardened cases than identical alloys induction heat treated under identical conditions, but containing an annealed microstructure. For the 1070 steel specimens, bars 2.54 cm (1 in.) in diameter were heated to 925 °C (1700 °F) in 1 s and water quenched. At this heating rate, the Ac_3 temperatures were found to be approximately 910 °C (1670 °F), 855 °C (1570 °F), and 815 °C (1500 °F) for the annealed, normalized, and quench-and-tempered microstructures, respectively. It

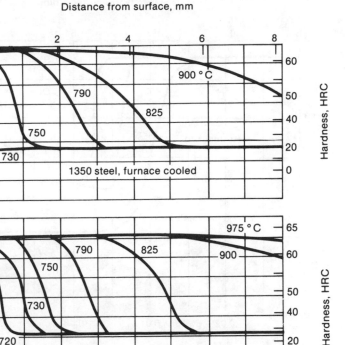

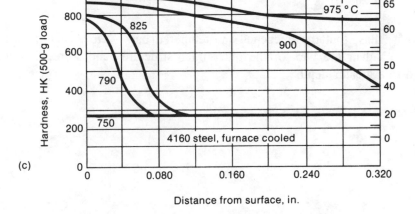

(a) 1350 steel. (b) 2350 steel. (c) 4160 steel. The bars were induction heated to the indicated maximum surface temperatures using a 530-kHz, 15-kW generator and then water quenched. The A_3 temperatures of the steels were (a) 750 °C (1380 °F); (b) 725 °C (1335 °F); and (c) 775 °C (1425 °F).

Fig. 5.24. Subsurface hardness distributions in steel bars 2.22 cm (0.88 in.) in diameter and 2.54 cm (1 in.) long, each with a furnace-cooled starting microstructure (Ref 56)

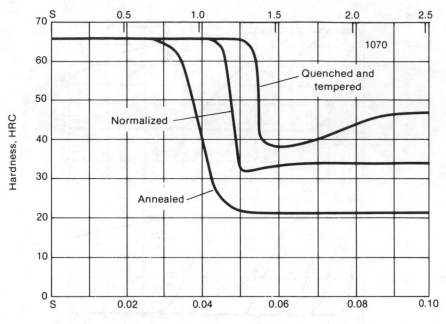

Fig. 5.25.   **Effect of starting microstructure in 1070 steel bars on response to surface hardening using a 450-kHz induction generator operated at a power density of 2.5 kW/cm² (15.9 kW/in.²) (Ref 50)**

is apparent, thus, that the quenched-and-tempered bars were austenitized to a greater depth than the bars of the other microstructures and were, therefore, hardened to a greater depth.

Chapter 6

Induction Tempering of Steel

Induction tempering, like hardening, has been found to be a viable commercial process, replacing conventional furnace operations in many high-production applications such as oil-well pipe and railroad rails. Unlike induction hardening, however, the controls needed for successful induction tempering must be more stringent in terms of both time and temperature. In hardening processes, austenitizing times and temperatures can be varied within wide limits as long as certain minimums are achieved, and identical hardening response will be obtained. By contrast, time and temperature are both critical in tempering. Increases in either of these may lead to substantial decreases in hardness. Thus, it is important that equivalent times and temperatures be established when induction tempering lines are set up initially or subsequently modified. The metallurgical correlations needed to accomplish this have been briefly described in Chapter 2. In the present chapter, this information will be reviewed. Attention will be focused on the AISI 10xx to 92xx series steels. Induction tempering of stainless steels and tool steels is discussed in Chapter 9. In addition, equipment used for induction tempering and typical properties of induction tempered products will be discussed.

METALLURGY OF SHORT-TIME TEMPERING

The basis for induction tempering lies in the fact that conventional long-time, low-temperature treatments can be replaced by short-time, higher-temperature ones. This is a result of the diffusion-controlled precipitation process taking place during tempering and the increase in the speed with which it occurs with increasing temperature. Data illustrating the effects of time and temperature are presented in Fig. 6.1 for a 1050 carbon steel austenitized and quenched to a martensitic microstructure with a hardness of 62 HRC. For a 1-h furnace heat treatment, the total amount of softening increases with temperature as shown. Similar responses are obtained for induction heat treatments for shorter times and higher temperatures. For instance, a 40 HRC temper is produced by a 1-h furnace heat treatment at 425 °C (800 °F) as well as by a 5-s induction treatment at 540 °C (1000 °F).

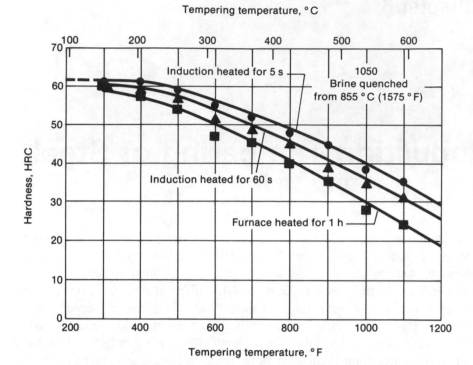

Fig. 6.1.   Hardness as a function of tempering temperature and time for furnace and induction treated 1050 steel austenitized at 860 °C (1575 °F) and quenched in brine (Ref 50)

Hollomon-Jaffe Tempering Correlation

As was discussed in Chapter 2, relations for the time-temperature equivalence of different treatments, such as those employing furnace and induction heating, were investigated first by Hollomon and Jaffe (Ref 27). According to their formulation, the tempered hardness of martensite is a unique function of the quantity $T(C + \log_{10} t)$, where T is absolute temperature, C is a material constant, and t is time in seconds.* In the above example, the constant C can be found by equating $(425 + 273)(C + \log 3600)$ to $(540 + 273)(C + \log 5)$, resulting in a value of C of approximately 17. With this value for C, equivalent time-temperature combinations for other hardnesses of the quenched 1050 steel can be obtained. For instance, a 30 HRC temper obtained in a 1-h furnace treatment at 540 °C (1000 °F) may be obtained by a 60-s induction treatment at a temperature of $(540 + 273)(17 + \log 3600)/(17 + \log 60) = 890$ K $= 615$ °C (1140 °F), which is in approximate agreement with the experimental data in Fig. 6.1.

The values of C in the Hollomon-Jaffe relation for several plain carbon steels are given in Table 6.1. From this listing, it can be seen that the magnitude of C

---

*In subsequent discussion, the base of the common logarithm, 10, will be omitted and "$\log_{10}$" will be written simply as "log".

Table 6.1. Hollomon-Jaffe C parameters for tempering of various plain carbon steels (Ref 27)

Steel	Starting microstructure	C
1030	Martensite	15.9
1055	Martensite	14.3
1074	Martensite	13.4
1090	Martensite	12.2
1095	Martensite	9.7
1095	Martensite + retained austenite	14.7
1095	Bainite	14.3
1095	Pearlite	14.1

tends to increase with decreasing carbon content. In general, C tends to lie between 14 and 18 for steels with carbon contents below 0.5%, and between 10 and 14 for steels of higher carbon content. Note that these values apply only when the tempering time t is expressed in seconds. If t is given in hours, the values of the constant should be increased by 3.6 ($\approx$ log 3600).

The values of C for plain carbon steels in Table 6.1 were taken from the work of Hollomon and Jaffe. As just mentioned, the values of this quantity appear to depend greatly on carbon content. From Table 6.1, a C value of approximately 14.5 would be estimated for 1050 steel. This is somewhat different than the value of 17.0 estimated from Fig. 6.1. For the two 40 HRC tempering treatments discussed above (1 h at 425 °C or 800 °F, and 5 s at 540 °C or 1000 °F), a correlation based on C = 17.0 gives a tempering parameter of about 14,370 in both cases. On the other hand, using a C value of 14.5 yields tempering parameters of 12,605 and 12,355 for the low- and high-temperature treatments, respectively. Inspection of curves such as the one in Fig. 2.22(a) reveals that such differences in the tempering parameter would give rise to differences in predicted hardness of perhaps one HRC point, or an amount well within experimental accuracy. Because of this, it can be concluded that the correlation of tempered hardness data is not especially sensitive to the value of C employed, within certain limits, of course. This hypothesis was verified by Grange and Baughman (Ref 28), who determined that a C value of 14.44, when t is expressed in seconds and T in degrees Rankine, results in very good hardness correlations for a large variety of carbon and alloy steels such as those in the AISI/SAE 10xx to 92xx series (Table 2.1). This work will be discussed later in this chapter.

Extension of Tempering Correlations to Continuous Heating/Cooling Cycles

The Hollomon-Jaffe equation, although quite useful in conjunction with conventional tempering curves, should be applied with care for induction tempering of martensite. Firstly, it must be remembered that there is a limit above which the tempering temperature should not be raised. This, of course, is the A_1 temperature (or Ac_1 for rapid heating processes), at which carbides start to go back into solution. Secondly, it must be realized that the relation applies only to short-time tempering at a *fixed* temperature, i.e., "isothermal" tempering treatments. In other words, it assumes that the temperature of the workpiece is increased instantaneously to the tempering temperature. When the heating time is of the same order of magnitude as the actual soak time, it must be taken into account.

A means by which a particular time-temperature history is accounted for in rapid heating (for instance, by induction) may be derived by a simple extension of the Hollomon-Jaffe concept. This is done by calculating the *equivalent* time t^* for a *constant* temperature heating cycle which corresponds to the continuous cycle. One way of doing this is illustrated in Fig. 6.2. Here, the induction tempering cycle (shown schematically in Fig. 6.2a) consists of a heating portion and a subsequent cooling portion, the latter occurring at a somewhat lower rate. The total continuous cycle is broken into a number of very small time increments, each of duration Δt_i and characterized by some average temperature T_i. It is assumed that the temperature for the equivalent isothermal treatment is the peak temperature of the continuous cycle, or T^*. This specification of the temperature for the isothermal cycle is arbitrary, however.

Having specified the temperature of the equivalent isothermal cycle as T^*, an effective tempering time t^* for this cycle can be estimated. This is accomplished by solving for the increment in t^*, or Δt_i^*, for each Δt_i in the continuous treatment by using the equation $T_i(C + \log \Delta t_i) = T^*(C + \log \Delta t_i^*)$. Summing the Δt_i^* for each portion of the continuous cycle yields the total effective tempering time t^* at temperature T^* and hence the "effective tempering parameter" $T^*(C + \log t^*)$, as shown in Fig. 6.2(b).

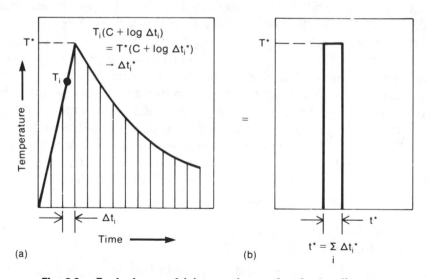

Fig. 6.2.   Equivalence of (a) a continuous heating/cooling tempering cycle and (b) an "isothermal" treatment through the use of an effective tempering time (t*) and temperature (T*) (Ref 68)

In the application of this method, care should be exercised in selecting Δt_i. These time increments should be chosen small enough so that the temperature does not change too much during the increment, thus enabling a reasonable average for the temperature T_i to be obtained and used in the above expression. For continuous heating from room temperature to typical induction tempering temperatures, a Δt_i on the order of 0.005 to 0.01 times t_{total}, where t_{total} is the total heating time, provides sufficient calculation accuracy.

Another consideration in estimating the effective tempering time is the fact that tempered steels are usually air cooled in order to avoid distortion. As implied above, the cooling rates are typically much lower than the heating rates, giving rise to substantially greater times at high temperature during the cooling cycle. Therefore, the tempering that occurs during cooling must also be included in the effective tempering parameter. To do this, the cooling rate must be measured or estimated from a heat-transfer analysis. The increments in the effective tempering time Δt_i^* are then estimated from this cooling curve and the above relation, and they are added to those for the heating portion of the cycle prior to calculation of the effective tempering parameter from $T^*(C + \log t^*)$.

Estimating Cooling Rates for Induction Tempered Steels

Following induction heating, steel parts lose heat to the surroundings by a combination of free convection and radiation. Books on heat transfer (e.g., Ref 69) supply formulae by which the net cooling rate can be estimated. Because induction finds the greatest application for round parts, the means by which cooling characteristics can be predicted in these cases will be discussed. For other geometries, the reader is referred to the standard texts.

The rate of heat loss of a round horizontal bar by free convection in air is best approximated by calculating the convective heat-transfer coefficient, h_m, which is a function of the nondimensional quantity known as the Nusselt number, Nu_m, which itself may be related to two other nondimensional numbers, the Grashof and Prandtl numbers, Gr and Pr, respectively:

$$Nu_m \approx 0.525(Gr \cdot Pr)^{0.25}$$

The product $Gr \cdot Pr$ for free convection in air is approximately equal to $15 \times 10^8 (D^3)(T_B - T_{air})/(T_B + T_{air})$, where D is the bar diameter in feet and T_B and T_{air} are the absolute temperatures of the bar surface and the ambient air in degrees Rankine ($°R = °F + 460$). The convective heat-transfer coefficient h_m is then given by $h_m = Nu_m k_{air}/D$, where k_{air} is the thermal conductivity of air (≈ 0.0152 Btu/h · ft · °R). Finally, the rate of convective heat loss, q_a, from a bar of length L is expressed simply by:

$$q_a = h_m A(T_B - T_{air}) = h_m \pi DL(T_B - T_{air})$$

where A is the surface area of the bar $(= \pi DL)$. Alternatively, the convective heat loss can be expressed as a rate per unit length, which from the above is obviously:

$$q_a/L = h_m \pi D(T_B - T_{air})$$

A simple expression for the rate of heat loss via radiation is also readily available. Neglecting the heat radiated from the surroundings *into* the bar, the radiation heat loss rate per unit length, q_r/L, is found from:

$$q_r/L = \pi De\sigma T_B^4$$

Here e is a number known as the emissivity, σ is the Stefan-Boltzmann constant $(= 1.712 \times 10^{-9} \text{ Btu/h} \cdot \text{ft}^2 \cdot °R^4 = 1.355 \times 10^{-12} \text{ cal/s} \cdot \text{cm}^2 \cdot K^4)$, and T_B is the absolute temperature of the bar again. The emissivity varies between 0 and 1. Lower values apply to bright, reflective surfaces, and higher values to dark surfaces. A value of 1 is used for so-called "black-body" radiation. For steels tempered around 540 °C (1000 °F), e values of approximately 0.5 to 0.75 pertain. In quick induction treatments, where little scaling occurs, values in the lower range are probably more applicable.

The net rate of heat loss due to convection and radiation is obtained by simply adding q_a/L and q_r/L. The *average* rate of cooling of the bar, $\Delta T/\Delta t$, is then given by:

$$\Delta T/\Delta t = [(q_a/L) + (q_r/L)]/A_x \rho_s c$$

where A_x is the cross-sectional area of the bar, ρ_s is the density, and c is the specific heat.

It should be realized that, in all of the above, a solid bar has been assumed. However, identical relationships hold for a tubular product provided that it is realized that D is the outer diameter of the bar and A_x is its cross-sectional area $[= 0.25\pi(OD^2 - ID^2)]$. In addition, end effects have been neglected. For long products, such as those that are processed in continuous heat treatment lines, these effects are most likely negligible.

Application of the Hollomon-Jaffe Correlation to Induction Tempering of Line Pipe

The application of the above relations will be illustrated with data on the induction tempering of a grade of tubular oil-country piping of the same nominal composition as that of 1030 steel, for which C = 15.9 (Table 6.1). This product has been processed in a continuous heat treatment line to produce a hardness of approximately 26 HRC (UTS = 870 MPa = 126 ksi) using the time-temperature cycle (for heating) shown in Fig. 6.3. The induction generator had a frequency of 300 Hz. This, combined with a relatively low power density and a long processing line consisting of eight heating stations, ensured relatively uniform heating through the 10-mm (0.42-in.) thick wall of the 140-mm (5.5-in.) OD pipe.

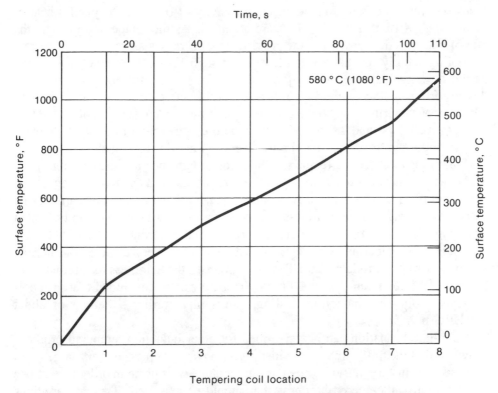

Time, s

Tempering coil location

Fig. 6.3. Typical heating cycle for induction tempered seamless pipe with a 140-mm (5.5-in.) OD and a 10-mm (0.42-in.) wall thickness (Ref 70)

The effective time t* for the heating portion of the tempering cycle was calculated as discussed previously. The temperature T* was taken to be equal to the peak temperature in the cycle, or 855.2 K (1080 °F). Furthermore, the cycle was broken into time increments of 0.5 s. Thus, the increments Δt_i^* were estimated from log Δt_i^* = $(T_i/855.2)(15.9 + \log 0.5) - 15.9$. For the time increment at which T_i = 800 K, Δt_i^* = 0.049 s. For the time increment at which T_i = 850 K, Δt_i^* = 0.402 s. Thus, it is apparent that relatively large contributions to the effective tempering time come from temperatures near the peak temperature, and relatively small amounts are a result of tempering that occurs at temperatures 50 K (90 °F) or more below the peak temperature. When all of the Δt_i^* values are summed, the total effective t* is found to be equal to 4.92 s, a time considerably shorter than the total heating time of 110 s. At the completion of tempering, the effective tempering parameter for this grade of pipe is therefore 855.2 (15.9 + log 4.92) = 14,190.

The tempering parameter is increased somewhat during slow cooling of the pipe in cooling beds following induction heating. Assuming cooling by free convection and radiation, the initial rate of cooling for the pipe from 855.2 K (1080 °F) is estimated to be 0.5 K/s (0.9 °F/s) using the relationships presented in the previous section. To obtain this value, an emissivity of 0.75 and a specific heat of 0.18 cal/g · K (0.18 Btu/1b · °F) were assumed. As the temperature drops, the rate of heat loss decreases, but the specific heat decreases as well. Thus, the rate of cooling is fairly constant over at least the first 50 K (90 °F) or so of the cooling period during which the vast majority of the softening resulting from the cool-down portion of the tempering occurs.* The effective time for the tempering which occurs during the cooling period, assuming a constant cooling rate of 0.5 K/s (0.9 °F/s) and $T^* = 855.2$ K as before, was found to be 41.57 s. This time is about ten times that for the much more rapid heating cycle.

Combining the effective tempering times for both the heating and cooling cycles leads to a t^* value of about 46.5 s for the entire process and an effective tempering parameter of 15,025, or a quantity about 800 greater than that for the heating cycle alone. An estimate of the hardness of the pipe after heat treatment can be gotten from Fig. 2.22(a). From this plot, it is found that a hardness of 25 HRC corresponds to a tempering parameter of 15,025. This is very close to the normal hardness for this product, 26 HRC. In contrast, the hardness is predicted to be 28.5 HRC for a parameter of 14,190. Thus, it can be concluded that the cooling cycle following induction heating results in a further softening of about 3.5 HRC points.

A similar calculation can be performed for other induction tempered products provided that the time-temperature history is known. For parts of very large cross sections, this history must be known throughout the workpiece in order to estimate the variations in tempered hardness that should be expected. Typical variations will be discussed later in this chapter.

Grange-Baughman Tempering Correlation

A correlation of constant-temperature ("isothermal") cycles for martensite tempering which is very similar to that of Hollomon and Jaffe is one which was established by Grange and Baughman (Ref 28). The form of the parameter in their work is the same, namely, $T(C + \log t)$, where T is the absolute tempering temperature, t is the tempering time in seconds, and C is a constant. Unlike the Hollomon-Jaffe formulation, however, C is a fixed number (equal to 14.44 when t is in seconds) irrespective of the alloy content. The success of the Grange-Baughman method is demonstrated by the results shown in Fig. 6.4 for a series of plain carbon steels. Note that the absolute temperature is in degrees Rankine, and not in kelvins, in this illustration. The fact that equally good correlation is obtained by both the Hollomon-Jaffe and the Grange-Baughman results again suggests that the exact value of C has little effect and that the most important variables are the absolute temperature and the logarithm of the tempering time.

Grange and Baughman also presented a means by which the tempered hardness in an *alloy* steel (10xx to 92xx series) could be estimated. This is done by first obtaining a "base" hardness from a tempering curve for the mild steel of the same

---

*Laboratory measurements and theoretical calculations confirmed this approximation as well as the fact that conduction effects minimize thermal gradients through the wall thickness.

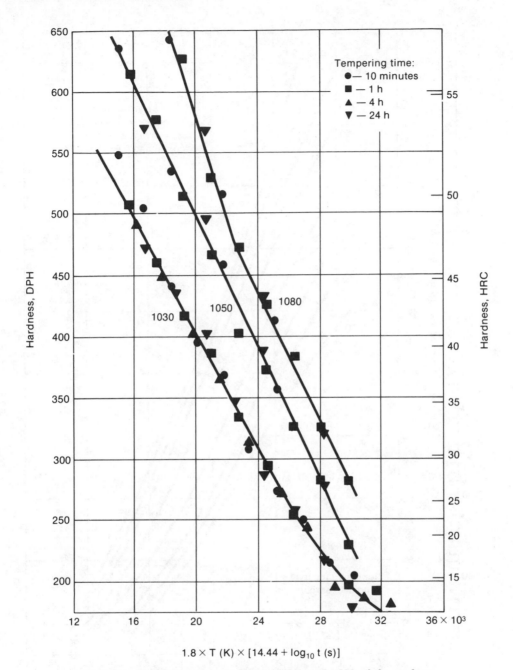

Fig. 6.4. Hardness data for 1030, 1050, and 1080 plain carbon steels plotted in terms of the Grange-Baughman tempering parameter (Ref 28)

carbon content (e.g., Fig. 6.5). To this are added certain hardness increments for each alloying element. These increments depend on the amount present and the value of the tempering parameter. The hardness increment (in DPH points) for each element is equal to the product of a certain factor (given in Table 6.2) and

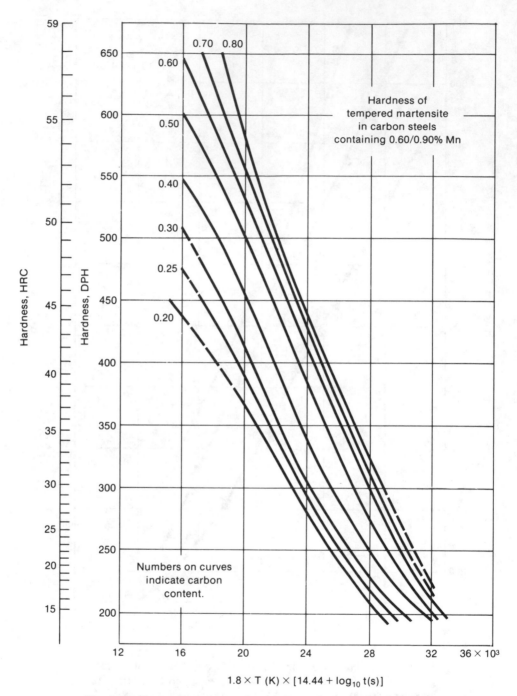

Fig. 6.5. Tempering curves for plain carbon steels plotted in terms of hardness as a function of the Grange-Baughman tempering parameter (Ref 28)

Table 6.2.   Hardness increment factors for various elements in tempered alloy steels (Ref 28)

Element(a)	Range, %	Hardness increment factor at parameter value of:					
		20,000	22,000	24,000	26,000	28,000	30,000
Manganese	0.85 to 2.1 35		25	30	30	30	25
Silicon	0.3 to 2.2 65		60	30	30	30	30
Nickel	Up to 4 5		3	6	8	8	6
Chromium	Up to 1.2 50		55	55	55	55	55
Molybdenum(b)	Up to 0.35 40		90	160	220	240	210
		(20)	(45)	(80)	(110)	(120)	(105)
Vanadium(c)	Up to 0.2 0		30	85	150	210	150

(a) Note: factor for boron is 0. (b) If 0.5 to 1.2% Cr is also present, use factors in parentheses. (c) For SAE/AISI chromium-vanadium steels; may not apply when vanadium is the only carbide-forming element present.

the percentage of that element minus some base percentage. For manganese and silicon, the base percentages are 0.85 and 0.3, respectively. For nickel, chromium, molybdenum, and vanadium they are zero.

A large amount of the tempering data used by Grange and Baughman in determining their tempering correlation was for treatments of 10 min or more. Only a limited amount was obtained for the shorter times typical of induction processes. However, work by Semiatin and his co-workers (Ref 68) verified the approach developed by Grange and Baughman for times as short as 6 s. To do this, samples 0.20 cm (0.080 in.) thick were quench hardened and tempered in salt pots. Tempering time was either 6, 60, or 600 s at temperature. Times to heat to tempering temperature were estimated using sheathed thermocouples of a diameter about equal to the sample thickness. By this means, it was determined that heating times were on the order of 3 to 4 s in all cases. Since they involve continuous heating, these times represent a rather small contribution to the effective tempering parameter, however.

Short-time tempering data obtained by Semiatin *et al* for 1020, 1042, and 1095 (whose chemistries are given in Table 6.3) are compared in Fig. 6.6 with "base" hardness data for 1020, 1050, and 1080 steels obtained by Grange and Baughman. For the 1042 and 1095 steels, the trends are very reasonable. However, short-time 1020 results diverge considerably from the 1020 base hardness curve. An explanation for this was found in a chemical analysis of the 1020 steel used in the short-time tempering trials. The analysis revealed (in weight percent) 0.22 C, 0.81 Mn, 0.18 Si, 0.014 P, 0.036 S, 0.13 Ni, 0.18 Cr, and 0.046 Mo. This chemistry is typical for 1020 except for the manganese, nickel, chromium, and molybdenum contents, whose presence suggests the use of alloy steel scrap during steel refining. From the data in Table 6.2, the increase in hardness due to these additions can be estimated. That due to manganese is zero because its concentration is below the "base" amount of 0.85 wt %. At a parameter level of 24,000, for instance, the increase for the other elements is $(0.13)(6) + (0.18)(55) +$

Table 6.3.　Chemical analyses of steels used in investigation of Semiatin, Stutz, and Byrer (Ref 68)

Steel	C	Mn	P	S	Si	Cu	Sn	Ni	Cr	Mo	Al	V	Co
1020...	0.22	0.81	0.014	0.036	0.18	0.17	0.010	0.13	0.18	0.046	0.003	0.001	0.005
1042...	0.44	0.92	0.025	0.050	0.26	0.029	0.003	0.053	0.078	0.019	0.039	...	0.002
1095...	0.96	0.45	0.023	0.029	0.24	0.013	0.002	0.021	0.094	0.015	0.025	0.002	...
4130...	0.32	0.52	0.012	0.021	0.25	0.11	0.007	0.13	1.04	0.16	0.024	0.005	0.005
4340...	0.40	0.76	0.008	0.020	0.28	0.13	0.009	1.62	0.85	0.22	0.039	0.001	0.010
4620...	0.17	0.54	0.007	0.016	0.29	0.17	0.009	1.80	0.14	0.26	0.012	0.001	0.033
8620...	0.19	0.83	0.016	0.025	0.25	0.054	0.004	0.48	0.56	0.19	0.041	0.001	0.006

Composition, wt %

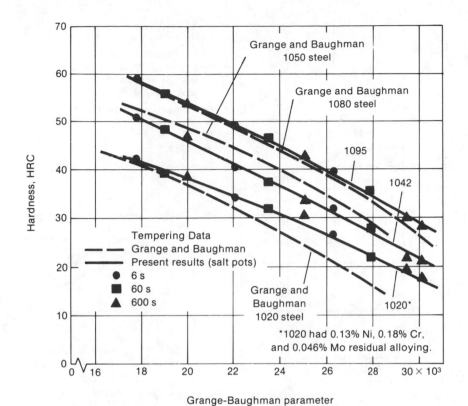

Fig. 6.6. Comparison of short-time tempering data for 1020, 1042, and 1095 carbon steels with "base" hardness curves for 1020, 1050, and 1080 steels from Grange and Baughman (Ref 68)

(0.046)(160) = 18 DPH points. This hardness increase should be added to the base hardness of 279 DPH (27 HRC) to obtain a predicted hardness of about 297 DPH, or 29.5 HRC. Other calculations of this type (at various parameter levels) result in the prediction in Fig. 6.7, which shows much better agreement with actual tempering data. Other short-time tempering data for 4130, 4340, 4620, and 8620 steels are depicted in Fig. 6.8. Agreement between the data and predictions based on the appropriate base hardness plots (Fig. 6.5) and the hardness increments (Table 6.2) is shown to be quite good.

Application of the Grange-Baughman Parameter to Induction Tempering

The Grange-Baughman correlation can also be applied to continuous heating cycles such as the tempering of the oil-country pipe product cited above. Complete chemical analysis of the steel in this case revealed, on the average, a composition (in weight percent) of 0.26 C, 0.23 Si, 1.31 Mn, 0.02 Cr, 0.02 Ni, 0.16 Mo, and 0.01 V. For the identical heat treatment and cooling cycle discussed above, the Grange-Baughman tempering parameter is found to be equal to 24,940. Re-

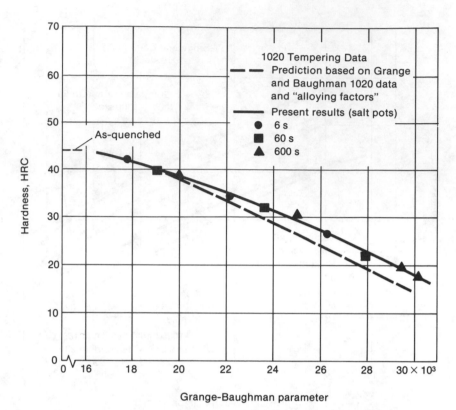

Fig. 6.7.   Comparison of short-time tempering data for 1020 steel (0.22C-0.81Mn-0.18Si-0.014P-0.036S-0.13Ni-0.18Cr-0.046Mo) with predictions based on Grange-Baughman "base" data for 1020 steel and hardness increment factors (Ref 68)

ferring to Fig. 6.5, and interpolating to 0.26% carbon, the base hardness of the equivalent carbon steel is estimated to be 24 HRC, or 257 DPH. At a tempering parameter level of 25,000, the increments in DPH points due to alloying are $(1.31 - 0.85)30 = 13.8$ for manganese, 0 for silicon, 1.1 for chromium, 0.1 for nickel, and 30.4 for molybdenum, resulting in a total predicted hardness increase of about 45 DPH points. Thus, the DPH hardness is estimated to be $257 + 45 = 302$, equal to an HRC hardness of about 30, which is 4 points higher than that ordinarily found for this grade and processing sequence. This deviation can be compared with that estimated from the Hollomon-Jaffe correlation (one point on the *low* side), which did not account for the different alloying elements present.

The induction tempering data obtained by Semiatin and his co-workers (Ref 68 and 71) add further credence to the concept of an effective tempering parameter based on the Grange-Baughman formulation. In this work, a number of austenitized-and-water-quenched bars of the steels listed in Table 6.3 were induction tempered. These bars were 2.54 cm (1.0 in.) in diameter and 15.24 cm

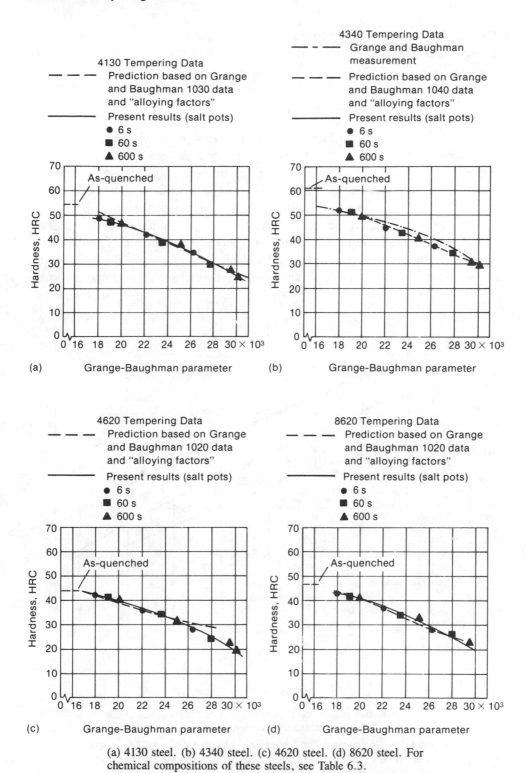

(a) 4130 steel. (b) 4340 steel. (c) 4620 steel. (d) 8620 steel. For chemical compositions of these steels, see Table 6.3.

Fig. 6.8. Comparison of short-time tempering data with predictions based on Grange-Baughman "base" hardness curves for carbon steels and hardness increment factors (Ref 68)

(6.0 in.) long; they were tempered using a 10-kHz motor-generator at a power setting of 5 kW. After being heated to a nominal peak surface temperature of 400, 540, or 675 °C (750, 1000, or 1250 °F), the bars were air cooled. Hardnesses were measured at the same near-surface location as that at which the temperature history had been measured for the purpose of calculating the effective tempering parameter.

Induction tempering data for the carbon steels listed in Table 6.3 (in terms of HRC hardness versus the effective Grange-Baughman parameter) are shown in Fig. 6.9. The over-all decrease in hardness with increasing tempering parameter seen in the isothermal salt-pot trials (Fig. 6.6) was replicated in the induction results. However, the induction samples exhibited somewhat lower hardnesses (1 to 3 HRC points) than the salt-pot samples, the measurements on which are indicated by the solid trend lines reproduced here from Fig. 6.6. The difference between the induction and salt-pot results was greatest for the 1020 steel. Nevertheless, when the 1020 induction data were compared with the tempering curve

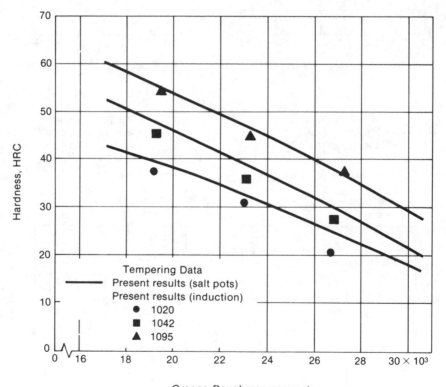

Grange-Baughman parameter

The induction results are plotted in terms of hardness versus the *effective* Grange-Baughman parameter.

Fig. 6.9. Comparison of tempering behavior of carbon steels heat treated in salt pots (solid lines) and by induction (data points) (Ref 68)

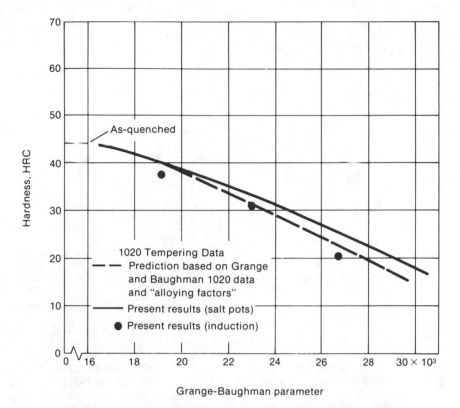

As-quenched

1020 Tempering Data
— — Prediction based on Grange
and Baughman 1020 data
and "alloying factors"
—— Present results (salt pots)
● Present results (induction)

Grange-Baughman parameter

Fig. 6.10. Comparison of tempering data for 1020 steel given salt-pot and induction treatments with a prediction based on Grange and Baughman's 1020 results and hardness increment factors ("alloying factors") (Ref 68)

prediction based on the Grange-Baughman 1020 steel and the hardness increment factors, the deviation was narrowed (Fig. 6.10). In fact, the prediction lay *between* the induction and the salt-pot results.

The induction tempering data for the alloy steels, whose chemistries are given in Table 6.3, revealed a trend similar to that for the carbon steels (Fig. 6.11). The hardnesses are lower than the corresponding salt-pot-treated samples (Fig. 6.8). For these steels, however, the difference is only about 1 to 2 hardness points, which is almost within experimental scatter.

From the above industrial and laboratory results, it can, therefore, be concluded that the effective tempering parameter offers a useful means of correlating induction temperature-time histories with mechanical properties such as hardness level.

EQUIPMENT FOR INDUCTION TEMPERING

The selection of equipment for tempering of steel using induction methods is similar to that for austenitizing. The major difference is that somewhat lower

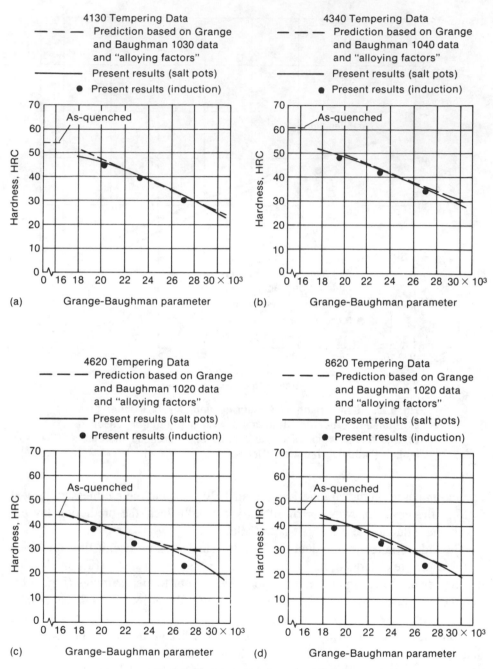

(a) 4130 steel. (b) 4340 steel. (c) 4620 steel. (d) 8620 steel. The induction results are plotted in terms of hardness versus the *effective* Grange-Baughman parameter.

Fig. 6.11.   Comparison of tempering data for alloy steels given salt-pot and induction treatments with predictions based on Grange and Baughman's carbon steel results and hardness increment factors ("alloying factors") (Ref 68)

frequencies are employed because the electrical and magnetic properties of steel are different at typical tempering temperatures. The resistivity ρ is lower, and the relative magnetic permeability is higher. The equation for reference depth reveals that, for temperature uniformity and efficiency equivalent to those for austenitizing applications, lower frequencies should be employed. Typical frequencies are given in Table 6.4 for "through"-tempering of steel bars of various sizes. As an example, the table shows that for tempering of stock 12.7 to 25.4 mm (0.5 to 1.0 in.) thick, a frequency of about 3000 Hz is best. From Table 5.2, the optimal frequency for through-hardening of stock in this diameter range is around 10,000 Hz.

As with through-hardening, the power density employed with induction tempering is kept very low, usually around 0.046 kW/cm^2 (0.3 kW/in.^{2}). As before, this is done to minimize the temperature gradients between the surface and center of the induction heated section. Typical operating parameters which satisfy this requirement are given in Table 6.5. Two examples from this tabulation will serve to show the kinds of temperature differentials and heating times that are representative of these situations. In both cases, electrical and radiation losses will be neglected.

As a first illustration, consider the induction tempering of a steel bar 12.7 mm (0.5 in.) in diameter from 50 to 565 °C (120 to 1050 °F) at a power rating of 11 kW and a power density of 0.064 kW/cm^2 (0.41 kW/in.^{2}). This is equivalent to an inductor whose length is 434 mm (17 in.). The power per unit length is, therefore, 0.647 kW/in. Also, using an a/d of 4, the induction thermal factor K_T is found to be 0.014 from Fig. 5.14 for the appropriate thermal conductivity of steel at the tempering temperature. The quotient of the power per unit length and K_T is equal to the "steady-state" temperature difference (in °F) between the surface and center of the bar. This difference is, therefore, 46 °F (26 °C). The heating *time* in this example can be found from the power rating and the heat content of a steel bar 12.7 mm (0.5 in.) in diameter and 434 mm (17 in.) long whose temperature is raised from 50 to 565 °C (120 to 1050 °F). From Fig. 5.11 and the weight of the bar, the heat supplied is equal to 0.038 kW · h or 136 kW · s, leading to a heating time of $136/11 \approx 12$ s. The actual heating time (17 s, from Table 6.5) is longer because of the electrical and radiation losses discussed in Chapter 5.

As a second example, consider the tempering of the 49-mm (1^{15}/$_{16}$-in.) diameter round bar in Table 6.5. In this case, the inductor length is equal to 500 mm (19.7 in.), and the power per unit length is 0.480 kW/cm (1.22 kW/in.). Using the same induction thermal factor as above (0.014), the temperature differential under steady-state conditions is 48 °C (87 °F).* This is about the upper limit on the temperature differential that should be acceptable for tempering. Even so, the actual heating time in this case (196 s) is long. Attempts at decreasing the tem-

---

*Note that during air cooling, heat conduction eliminates a large part of the thermal gradient, thus minimizing the gradient in properties of the final tempered product. This is discussed more fully at the end of this chapter.

Table 6.4. Selection of power source and frequency for various applications of induction tempering (Ref 50)

| Section size | | Maximum tempering temperature | | Rating(a) for: | | | | | |
cm	in.	°C	°F	Power lines, 50 or 60 Hz	Frequency converter, 180 Hz	Solid-state systems or motor-generators 1000 Hz	3000 Hz	10,000 Hz	Vacuum tube, over 200 kHz
0.32 to 0.64	1/8 to 1/4	705	1300	...	...	...	...	...	Good
0.64 to 1.27	1/4 to 1/2	705	1300	...	...	...	...	Good	Good
1.27 to 2.54	1/2 to 1	425	800	...	Fair	Good	Good	Good	Fair
		705	1300	...	Poor	Fair	Good	Good	Fair
2.54 to 5.08	1 to 2	425	800	Fair	Fair	Good	Good	Fair	Poor
		705	1300	...	Fair	Good	Good	Fair	Poor
5.08 to 15.24	2 to 6	425	800	Good	Good	Good	Fair	...	...
		705	1300	Good	Good	Good	Fair	...	...
Over 15.24	Over 6	705	1300	Good	Good	Good	Fair	...	...

(a) Efficiency, capital cost, and uniformity of heating are considered in these ratings. "Good" indicates optimum frequency. "Fair" indicates a frequency higher than optimum that increases capital cost and reduces uniformity of heating, thus requiring lower heat inputs. "Poor" indicates a frequency substantially higher than optimum that substantially increases capital cost and reduces uniformity of heating, thus requiring substantially lower heat inputs.

Table 6.5. Operating and production data for progressive induction tempering (Ref 50)

Section size cm	in.	Material	Frequency, Hz	Power(a), kW	Total heating time, s	Scan time s/cm	s/in.	Entering coil °C	°F	Leaving coil °C	°F	Production rate kg/h	lb/h	Inductor input(b) kW/cm²	kW/in.²
Rounds															
1.27	1/2	4130	9600	11	17	0.39	1	50	120	565	1050	92	202	0.064	0.41
1.91	3/4	1035 mod	9600	12.7	30.6	0.71	1.8	50	120	510	950	113	250	0.050	0.32
2.54	1	1041	9600	18.7	44.2	1.02	2.6	50	120	565	1050	141	311	0.054	0.35
2.86	1 1/8	1041	9600	20.6	51	1.18	3.0	50	120	565	1050	153	338	0.053	0.34
4.92	1 15/16	14B35H	180	24	196	2.76	7.0	50	120	565	1050	195	429	0.031	0.20
Flats															
1.59	5/8	1038	60	88	123	0.59	1.5	40	100	290	550	1449	3194	0.014	0.089
1.91	3/4	1038	60	100	164	0.79	2.0	40	100	315	600	1576	3474	0.013	0.081
2.22	7/8	1043	60	98	312	1.50	3.8	40	100	290	550	1609	3548	0.008	0.050
2.54	1	1043	60	85	254	1.22	3.1	40	100	290	550	1365	3009	0.011	0.068
2.86	1 1/8	1043	60	90	328	1.57	4.0	40	100	290	550	1483	3269	0.009	0.060
Irregular shapes															
1.75 to 3.33	11/16 to 1 5/16	1037 mod	9600	192	64.8	0.94	2.4	65	150	550	1020	2211	4875	0.043	0.28
1.75 to 2.86	11/16 to 1 1/8	1037 mod	9600	154	46	0.67	1.7	65	150	425	800	2276	5019	0.040	0.26

(a) Power transmitted by the inductor at the operating frequency indicated. For converted frequencies, this power is approximately 25% less than the power input to the machine, because of losses within the machine. (b) At the operating frequency of the inductor.

perature differential might involve lower power densities. However, this would further increase the heating time.

Other equipment for induction tempering is similar to that for induction hardening. This includes coil design, equipment for part rotation during heat treatment (to maintain temperature uniformity and minimize part distortion), and electronic controls which find particular importance in continuous heat treating lines.

PROPERTIES OF INDUCTION TEMPERED STEELS

Although relatively uniform temperatures and, thus, uniform properties through the section are obtainable in induction tempering of tubular products, the examples in the previous section show that difficulties may be encountered when solid sections are involved. An important question, therefore, is how large a property variation — or, more specifically, hardness variation — should one expect from typical induction tempering systems in these situations. To a first order, this can be estimated by comparing the Hollomon-Jaffe parameters for the surface and center temperatures.* Neglecting the log t terms, which are small relative to the value of C in induction treatments anyway, the ratio is equal to $(T + 55)(15.9) / T(15.9)$ for a 55 K (55 °C or 100 °F) temperature differential in a 1030 steel bar. For $T = 840$ K, this is equal to approximately 14,200/13,400. Referring to Fig. 2.22, a hardness difference of only about 3.5 HRC points $(32 - 28.5)$ would be expected in this and similar situations.

Data on induction tempered solid bars verify the above conclusion and lend support to the usefulness of induction heating for tempering of solid sections. Several examples are shown in Fig. 6.12 for steels induction hardened and tempered using typical processing parameters such as those in Table 6.5. The as-quenched hardness distributions, as well as those after tempering, are depicted. In the first two examples for 1037 and 1041 steels, the as-quenched hardness pattern shows a noticeable decrease (of 8 to 10 HRC points) away from the surface, a trend which is most likely attributable to a lack of adequate hardenability or an improper quenching technique. Fortunately, however, the hardness patterns in the same two steels following tempering are much more uniform, the average HRC variation being only 3 points.

This is readily explained by the fact that the hardnesses of tempered martensite, tempered pearlite, and tempered bainite for equivalent heat treatment cycles are similar, especially those involving longer times or higher temperatures which produce moderate to low hardnesses. An illustration of this behavior is shown in Fig. 2.25 in which the Hollomon-Jaffe correlation for various starting, or as-quenched, microstructures for a 1095 steel are given. In this instance, the initial hardnesses varied between 46 and 63 HRC. In spite of the differences in microstructure and starting hardness, the C constant was nearly identical in all cases;

---

*This, of course, does not take into account the smoothing of the thermal gradient by conduction following the heating cycle.

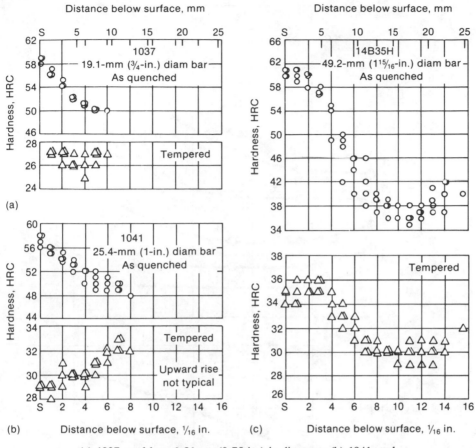

(a) 1037 steel bars 1.91 cm (0.75 in.) in diameter. (b) 1041 steel bars 2.54 cm (1.00 in.) in diameter. (c) 14B35H steel bars 4.92 cm (1.94 in.) in diameter.

Fig. 6.12. Hardness profiles in induction hardened and tempered steel bars (Ref 50)

and, for hardness levels of 40 HRC or lower, the tempered hardnesses were nearly equal for equal tempering temperatures and times.

The third steel, for which data are presented in Fig. 6.12, exhibits a similar *decrease* in hardness variation following tempering as compared with that from the hardening operation. After hardening, the 14B35H steel revealed a drop in hardness of approximately 22 HRC points from surface to center. This is most likely a result of low hardenability even though the steel does have a boron addition which tends to improve its ability to harden above that of ordinary plain carbon steels such as those depicted with the 14B35H steel in Fig. 6.12. Nevertheless, after tempering, the hardness variation is reduced from 22 points to 5 points.

Data related to those above are shown in Fig. 6.13 and 6.14. These results were obtained in the Soviet Union (Ref 72) on steels similar to AISI grades 1041 and 8740 and are for quench-hardened 168-mm (6.61-in.) OD, 145-mm (5.71-in.) ID

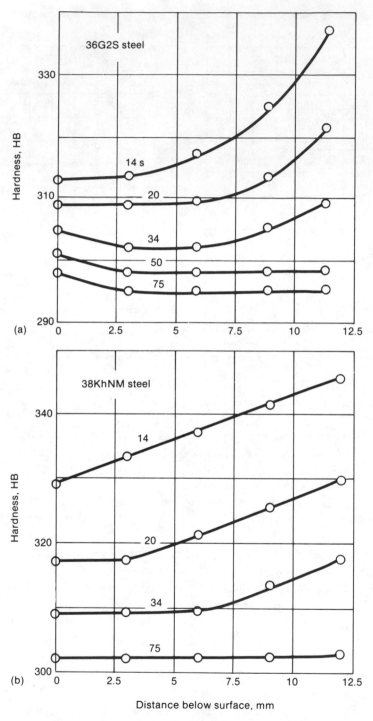

(a) 36G2S steel (≈ 1041). (b) 38KhNM steel (≈ 8740). The piping had an OD of 168 mm (6.61 in.) and an ID of 145 mm (5.71 in.).

Fig. 6.13. Hardness profiles through the wall of steel piping induction tempered using heating cycles lasting for the times indicated (Ref 72)

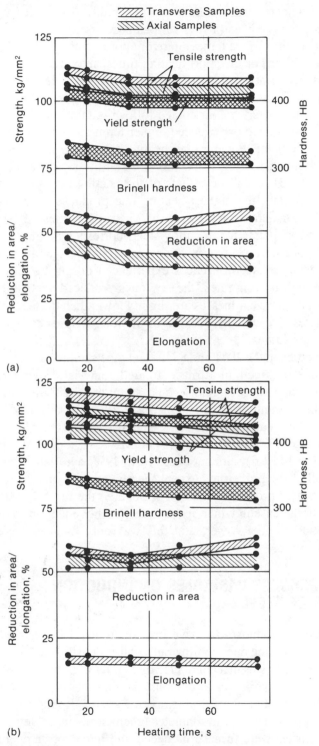

(a) 36G2S steel (≈ 1041). (b) 38KhNM steel (≈ 8740).

Fig. 6.14. Mechanical properties of induction tempered steel piping as a function of duration of heat treatment cycle (Ref 72)

piping induction heated using a 2500-Hz generator. Test pieces were induction scanned at various speeds and using various power settings such that the time to heat to the maximum surface temperature of approximately 650 °C (1200 °F) was between 14 and 75 s. For the short heating times, maximum temperature differences between the surface and interior layers of the pipe of about 100 °C (180 °F) were noted. The effects of such a large difference may have been minimized by the fact that test samples were air cooled following induction heating, and, thus, the inner layers may have experienced higher temperatures due to heat absorbed from the outer diameter. For the long heating times, the maximum temperature difference measured was only 10 °C (20 °F). Hardness profiles through the wall thickness (Fig. 6.13) reveal the effects of the heating rate and temperature differential. As expected, the highest heating rate produced hardness gradients in both steels. This gradient is on the order of 25 Brinell harness (HB) points for the carbon steel and 15 points for the alloy steel, corresponding approximately to 3 and 1.5 HRC points, respectively. Such variations appear rather minimal with regard to the temperature differences experienced during the heating cycle. In contrast to the high heating rates, the low rates produced very uniform hardness profiles. Over-all, these hardnesses are slightly lower than those resulting from the high heating rates, presumably because of the longer times at high temperatures for the low heating rates.

Tensile properties of the 1041 and 8740 steel pipe samples were found to be very similar to those obtained in furnace treated material. Specifically, furnace quenched and tempered 1041 had a yield strength of 788 to 832 MPa (114 to 121 ksi), tensile strength of 891 to 940 MPa (129 to 136 ksi), elongation of 15 to 18%, and reduction in area of 50 to 55%. The furnace treated 8740 steel exhibited a yield strength of 832 to 881 MPa (121 to 128 ksi), tensile strength of 930 to 979 MPa (135 to 142 ksi), elongation of 16 to 19%, and reduction in area of 50 to 55%. For all of the properties, except the reduction in area for the 1041 steel, the induction heated properties equaled or bettered the furnace treated properties (Fig. 6.14). For the case in which it was inferior, the reduction-in-area specification was only 10 to 15% lower, and this was only for samples cut from one direction in the pipe.

EFFECT OF PROCESS VARIABLES ON INDUCTION TEMPERING OF STEEL

Probably one of the most extensive investigations of the effects of induction heating variables (heating rate, cooling rate, power density, etc.) and hardenability on tempering response was that conducted by Semiatin, Stutz, and Byrer (Ref 71). The objective of this work was to demonstrate that relatively uniform properties could be obtained by short induction tempering cycles. Round bar samples which were either 2.54 cm (1.0 in.) in diameter (chemistries in Table 6.3) or 6.35 cm (2.5 in.) in diameter were furnace hardened and induction tempered. A limited number of tubes which were 6.35 cm (2.5 in.) in OD and 5.1 cm (2.0 in.) in ID

Table 6.6. Chemical compositions of 6.35-cm (2.5-in.) OD steel bars and tubes used in investigation of Semiatin, Stutz, and Byrer (Ref 71)

Steel	Product form	Composition, wt %								
		C	Mn	S	P	Si	Cu	Ni	Cr	Mo
1042	Bar	0.46	0.74	0.014	0.029	. . .	. . .	. . .	. . .	. . .
4340	Bar	0.40	0.76	0.015	0.025	0.27	0.12	1.84	0.84	0.23
8620	Tube . . .	0.18	0.74	0.027	0.005	0.20	0.12	0.50	0.49	0.18

were also heat treated in the study. The chemistries of the larger-diameter bars and the tubes are given in Table 6.6. The results of this investigation are described below.

Effect of Heating Rate. The effect of heating rate on induction tempering response was established by heating hardened 2.54-cm (1.0-in.) diameter samples of the steels listed in Table 6.3 using power settings of 2.5, 5.0, and 10 kW. These power settings gave rise to heating rates between 3 and 14 °C/s (5 and 25 °F/s). In all cases, bars 15.2 cm (6.0 in.) long were heated to a nominal peak surface temperature of 540 °C (1000 °F) and air cooled. Actual measured peak temperatures were between 525 and 550 °C (980 and 1020 °F).

In these experiments, temperature-versus-time histories were measured at the same near-surface location as that at which the hardnesses were taken. With these temperature measurements, effective tempering parameters (of the Grange-Baughman form) were estimated as described earlier in this chapter. These calculations revealed the parameter to vary only between approximately 22,500 (10-kW power setting) and 23,500 (2.5-kW power setting). The rather small variations in the effective tempering parameter in these cases can be ascribed to the nearly identical (air) *cooling cycle* in all experiments. Since the cooling rate was considerably lower than any of the heating rates, it is apparent that the cooling portion made a larger contribution to the effective tempering time and, thus, the effective tempering parameter.

Examination of tempering data, such as those shown in Fig. 6.6, shows that a variation of ±500 in the tempering parameter should give rise to rather small hardness variations (on the order of ±1 HRC point). This hypothesis was verified by the hardness measurements given in Table 6.7. It can be seen that, for each steel, the hardnesses were nearly independent of the heating rate (i.e., power setting). If there is a definite trend, the results suggested a very slight *increase* in hardness with heating rate. This trend is as expected in view of the slight decrease in the effective tempering parameter with increasing heating rate.

Effect of Cooling Rate. The influence of cooling rate following induction heating on tempering response was also explained by Semiatin *et al* through the application of the effective tempering parameter. This was done for a series of 25.4-cm (10-in.) long tubes of 8620 steel which were tempered by heating to nominal peak temperatures of either 540 or 675 °C (1000 or 1250 °F) and subsequently air cooled

Table 6.7. Effect of heating rate (power input) on hardness of induction tempered steels (Ref 71)

| | Hardness, HRC, at power input of: | | |
Alloy	2.5 kW	5 kW	10 kW
1020	30.2	30.7	31.3
1042	35.0	35.6	36.6
1095	44.5	44.6	46.4
4130	38.4	39.0	40.0
4340	41.4	41.8	41.4
4620	30.6	31.9	31.7
8620	32.2	32.8	33.0

Table 6.8. Effect of cooling method on hardness of 8620 steel tubes following induction tempering (Ref 71)

| Peak temperature | | Effective tempering time, s | Power setting, kW | Air cool (AC) or water quench (WQ) | Grange-Baughman effective tempering parameter | Hardness, HRC |
°C	°F					
535	995	35	5.0	AC	23,250	32.8
548	1018	9	5.0	WQ	22,720	33.8
681	1257	38	5.0	AC	27,500	22.0
678	1252	17	5.0	WQ	26,810	24.0
535	995	30	10.0	AC	23,150	32.9
563	1045	4	10.0	WQ	22,630	34.4
691	1275	31	10.0	AC	27,640	22.8
704	1300	6	10.0	WQ	26,770	25.4

or water quenched. Heating was done at a power setting of either 5 or 10 kW. The process parameters and calculated effective tempering parameters for this series of trials are summarized in Table 6.8.

For a given power setting and peak temperature, the results in Table 6.8 demonstrate that the major effect of cooling rate on the effective tempering characteristics was to reduce the effective tempering time. For example, for the experiments in which tubes were heated to approximately 540 °C (1000 °F), using a power setting of 5 kW, air cooling and water quenching led to effective tempering parameters of 23,250 and 22,720, respectively. This small parameter difference, despite the large effective tempering-time difference (35 versus 9 s), is a result of the fact that the time variable (t) enters the tempering parameter equation as log t. Also, the log t term is additive to a rather large number (14.44). The variation of tempering parameter with cooling method for the other pairs of data in Table 6.8 is similar.

The variation of tempering parameter with cooling method following induction heating correlated well with the measured mid-wall thickness hardness values

(Table 6.8). As expected, the air-cooled specimens exhibited hardnesses 1 to 2.5 HRC points lower than their water-quenched counterparts. The data in Table 6.8 also add further confirmation to the above conclusions regarding the effect of heating rate on induction tempering behavior. If the results for just air-cooled (or water-quenched) specimens heated to the same nominal peak temperature, but at different power settings, are compared, the hardnesses are seen to be nearly identical.

Effect of Temperature Gradients on Tempering Parameter Gradients. The major effect of induction heating on hardness and property gradients in through-tempered parts arises from the variation of the temperature-versus-time history experienced at various locations in the cross section. In addition, the existence of nonuniform starting or as-quenched through-thickness hardness patterns plays an important role. The influence of induction heating temperature gradients is discussed in this section. That of nonuniform initial hardness is addressed in the next section.

Induction heating of round bars (and parts of similar geometry) gives rise to distinctive heating patterns as discussed briefly in Chapter 5. Typical examples from the work of Semiatin *et al* are depicted in Fig. 6.15 and 6.16 for bars 2.54 and 6.35 cm (1.0 and 2.5 in.) in diameter heated to a nominal peak temperature of 705 °C (1300 °F). In both figures, the measured surface and center temperature histories for the induction heated solid round bars are plotted. The important features of these graphs are the following:

- The surface temperature exceeds the center temperature during the heating portion of the cycle.
- Beyond the initial heating transient, a steady-state temperature difference ΔT is developed. The magnitude of ΔT increases with the power setting.
- At the cessation of heating, the surface temperature begins to drop immediately (due to convection heat losses to the surroundings and conduction heat losses to the center of the workpiece). In contrast, the center temperature continues to rise. Moreover, the temperature maximum at this location is considerably more diffuse than that at the surface.
- Eventually, the center temperature slightly *exceeds* the surface temperature, and both positions cool at approximately the same rate thereafter.

The above characteristics play an important role in determining the difference in the effective tempering parameter between the center and surface of induction tempered bars. First, it is important to note that the difference in *peak* temperature between the two locations is considerably *less* than the steady-state ΔT, as the data in Table 6.9, which have been extracted from Fig. 6.15 and 6.16, reveal. This peak temperature difference is only about one-half the steady-state ΔT. In addition, the diffuseness of the center temperature history gives rise to somewhat larger effective tempering times. Both of these traits tend to minimize the variation in effective tempering parameter between the surface and center locations. For the

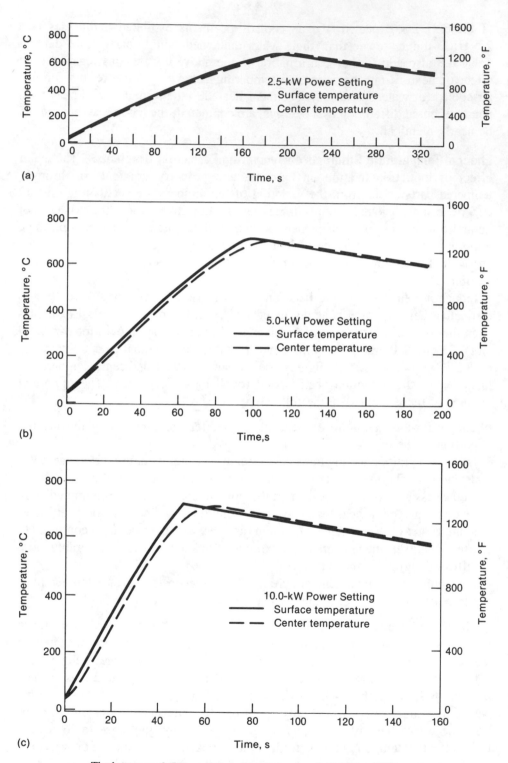

The bars were 2.54 cm (1.0 in.) in diameter and 15.1 cm (6.0 in.)
long. Power settings of (a) 2.5 kW, (b) 5 kW, and (c) 10 kW were
used during induction heating.

**Fig. 6.15.   Temperature-versus-time histories measured at the
surface (solid lines) and center (dashed lines) of induction
heated and air cooled bars of 1042 steel (Ref 71)**

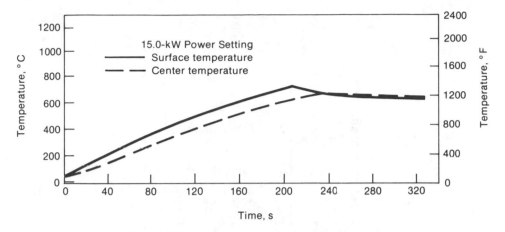

The bar was 6.35 cm (2.5 in.) in diameter and 25.4 cm (10.0 in.)
long. A power setting of 15 kW was used during induction heating.

**Fig. 6.16.   Temperature-versus-time histories measured at the
surface (solid line) and center (dashed line) of an induction
heated and air cooled bar of 1042 steel (Ref 71)**

heating trials listed in Table 6.9, these variations were less than 300 for 2.54-cm
(1.0-in.) diameter bars and less than 1000 for 6.35-cm (2.5-in.) diameter bars.
Based on the results shown in Fig. 6.6, such variations in effective tempering
parameter can be expected to produce maximum hardness variations of 2 HRC
points across the cross section of uniformly through-hardened steel bars.

Of the two major factors controlling the tempering parameter variation, i.e., the
variations in peak temperature and effective tempering time, the former is surely
the more important. This is because temperature enters the parameter equation
$[T*(C + \log t*)]$ as a multiplicative rather than an additive term. For this reason,
gross estimates of the difference in the tempering parameter and, thus, predicted
hardnesses based on steady-state ΔT's can be substantially greater than those
based on the peak temperatures at the surface and center.

Effect of Initial Hardness Pattern on Tempered Hardness Gradients.   The above
results establish that for the heating rates used in the experiments of Semiatin *et
al*, the maximum variation in tempered hardness from surface to center of the
2.54-cm (1.0-in.) or 6.35-cm (2.50-in.) diameter bars should be 2 HRC points.
Of course, this assumes a uniform starting, or as-quenched, hardness. The hardness uniformity in this condition depends on hardenability.

Of the alloys used in the investigation of Semiatin *et al*, 4130 and 4340 had the
highest hardenability as determined by estimates of the ideal critical diameter.
These calculations suggested that 2.54-cm (1.0-in.) diameter bars of both steels
and 6.35-cm (2.50-in.) diameter bars of 4340 could be readily through-hardened
to 100% martensite. Hardness measurements on as-quenched, cross-sectioned
samples supported this conclusion. For the 4130, a uniform hardness of 52 HRC
was obtained. The hardness for 4340 bars of both diameters was also fairly
uniform; it was 59 HRC. Based on these measurements and the above tempering

Table 6.9.   Effect of temperature gradients on tempering parameter gradients during induction heating (Ref 71)

| Bar diameter | | Power setting, kW | Steady-state ΔT(a) | | Location | Peak temperature | | Effective tempering time, s | Effective tempering parameter |
cm	in.		°C	°F		°C	°F		
2.54	1.0	2.5	14	25	Surface	704	1300	35	28,140
					Center	693	1280	43	27,980
2.54	1.0	5.0	28	50	Surface	710	1310	28	28,120
					Center	699	1290	36	28,000
2.54	1.0	10.0	56	100	Surface	718	1325	25	28,270
					Center	702	1295	34	28,020
6.35	2.5	15.0	97	175	Surface	721	1330	36	28,640
					Center	671	1240	74	27,730

(a) Steady-state $\Delta T = T_{surface} - T_{center}$ during steady-state *heating* portion of induction cycle.

Table 6.10.   Hardness measurements on induction tempered bars of high-hardenability steels (Ref 71)

Alloy	Bar diameter cm	in.	Power setting, kW	Peak surface temperature °C	°F	Hardness, HRC, at: Surface	Center
4130	2.54	1.0	2.5	535	995	38.4	38.3
			5.0	424	796	44.3	43.9
			5.0	545	1013	39.0	38.9
			5.0	677	1250	30.1	30.2
			10.0	531	987	40.0	39.9
4340	2.54	1.0	2.5	546	1014	41.4	42.0
			5.0	413	776	48.0	48.7
			5.0	540	1004	41.8	42.0
			5.0	677	1250	34.3	35.1
			10.0	553	1027	41.4	42.3
4340	6.35	2.50	5.0	539	1002	40.4	40.9
			10.0	685	1265	33.3	33.4
			15.0	543	1010	42.3	43.4

parameter calculations, the *tempered* hardnesses of bars of these two steels would be expected to be fairly uniform. This conclusion was supported by the data in Table 6.10. Experimental scatter of ±1 HRC point tends to mask any effect of tempering parameter variations on hardness gradients across the diameter, which would be very small anyway (a maximum gradient of 2 HRC points).

In contrast to 4130 and 4340, the other steels studied by Semiatin *et al* (1020, 1042, 1095, 4620, and 8620) had significantly lower hardenability. This fact was manifested by as-quenched hardness patterns measured across 2.54-cm (1.0-in.) diameter bars of these steels (Fig. 6.17). The *surface* hardnesses for these steels were indeed representative of structures of 100% martensite, but lower subsurface hardnesses (and accompanying metallography) established a lack of full hardening.

The hardness gradients shown in Fig. 6.17 for the low-hardenability steels were carried over, to a certain degree, in the induction tempered hardness distributions. Sample tempered hardness measurements from these experiments are given in Table 6.11.

The data in Table 6.11 show that, in some instances, the hardness gradient has been substantially decreased or almost eliminated (e.g., the data for 1095, 4620, and 8620). This behavior can be rationalized by reference to Fig. 6.18, a schematic illustration of the tempering response of starting microstructures of martensite, bainite, and pearlite; the basis for these curves is found in the research of Hollomon and Jaffe (Fig. 2.25), which was discussed earlier in this chapter and in Chapter 2. In the work of Semiatin *et al*, the microstructures in the as-quenched bars consisted actually of combinations of the three constituents shown in Fig. 6.18. However, it was concluded that tempering behavior at various locations

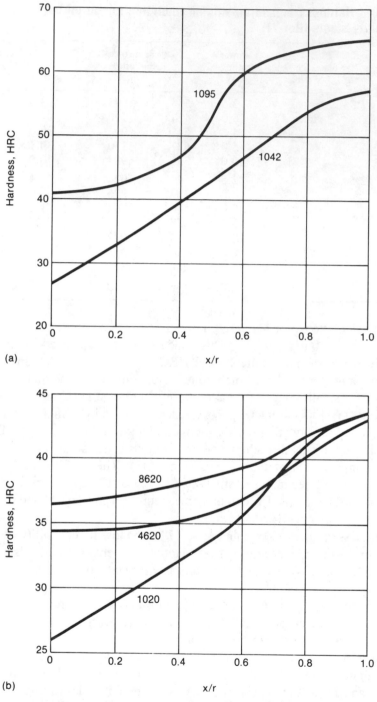

(a) 1042 and 1095 steels. (b) 1020, 4620, and 8620 steels. The hardnesses are plotted versus the fractional radial distance (x/r) from the center of the bar $(r = 0)$.

Fig. 6.17.   Rockwell C hardness profiles in 2.54-cm (1.0-in.) diameter steel bars following furnace austenitizing and water quenching (Ref 71)

Table 6.11.   Hardness measurements on induction tempered bars of low-hardenability steels (Ref 71)

All samples were heated to a nominal peak surface temperature of 540 °C (1000 °F) at a power setting of either 5 kW (2.54-cm-diam bars) or 15 kW (6.35-cm-diam bars). In all cases, the effective tempering parameter at the surface and center locations was 23,000 ± 600.

Alloy	Bar diameter		Hardness, HRC, at:	
	cm	in.	Surface	Center
1020	2.54	1.0	30.7	23.7
1042	2.54	1.0	35.6	27.5
	6.35	2.5	36.1	24.3
1095	2.54	1.0	44.6	43.3
4620	2.54	1.0	31.9	28.8
8620	2.54	1.0	32.8	30.6

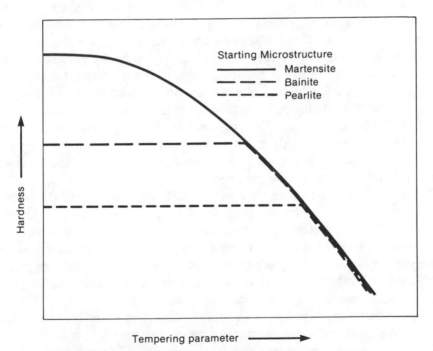

Fig. 6.18.   Schematic illustration of the tempering behavior (in terms of hardness versus tempering parameter) of a hypothetical steel initially heat treated to produce a microstructure of martensite, bainite, or pearlite (after Ref 27)

could still be represented by curves such as these; the lower curves would be used for the interior locations and the upper curve for surface and near-surface locations.

Examination of Fig. 6.18 reveals that, to eliminate hardness gradients in bars of low hardenability, large values of the tempering parameter must be employed. Specifically, the tempering parameter must be large enough so that the hardness of the tempered martensite is dropped at least to the level of the as-quenched hardness of the lower-temperature transformation product. With this proviso, the results in Table 6.11 were easily explained. For the 1020 and 1042 steels, tempering gave rise to surface hardnesses *above* those of the as-quenched structures at the centers of the bars, thereby leaving these locations relatively untempered. On the other hand, the 1095, 4620, and 8620 steels were tempered to give hardnesses comparable to, or lower than, the corresponding as-quenched hardnesses at the centers of bars of these steels. Neglecting the small differences in the effective tempering parameter discussed in the previous section, it is, therefore, not unexpected that the hardness gradient in the latter three steels was eliminated, at least to a first order.

Conclusions. From the work of Semiatin, Stutz, and Byrer, the following conclusions regarding tempering by induction methods can be drawn:

- The heating rate during induction tempering can be varied over a wide range without noticeably affecting the final tempered hardness. This is a result of the slow air cooling which typically follows heat treatment. Because cooling occurs at a much lower rate than heating, the contribution of the cooling cycle to the effective tempering time and, hence, effective tempering parameter is ordinarily larger than that of the heating portion of the tempering treatment.
- Similarly, for a given heating rate, the method of cooling has a second-order effect on the effective tempering parameter and, thus, tempered hardness levels. This behavior is a consequence of the fact that time enters the tempering parameter equation as a logarithm whose magnitude for induction tempering processes is typically much less than the constant to which this function is added when deriving the tempering parameter.
- The effect of surface-to-center temperature differences during induction heating of solid cross sections for the purpose of tempering is mitigated, to a large extent, by the conduction of heat into the interior after the power is turned off. Because of heat conduction, the interior temperature continues to increase while the surface is cooling, leading to differences between the peak temperatures experienced by the surface and center which are considerably smaller than the steady-state temperature difference during the heating cycle. In addition, the effective tempering time for the central regions of an induction heated bar are somewhat greater. These characteristics of the heating process minimize through-thickness variations in effective tempering parameter. Thus, they enable induction heating to be applied to impart uniform tempered hardnesses across sections which have been previously austenitized and quenched to a uniform hardness.
- Induction heating may also be used to through-temper cross sections which have not been previously hardened uniformly. Such tempering treatments tend to decrease pre-existing hardness gradients by an amount which increases with the effective tempering parameter. As-quenched hardness gradients can be essentially eliminated in induction tempering processes in which the effective tempering parameter equals or exceeds the value at which the *tempered surface hardness* equals the *as-quenched center hardness* of the part.

Chapter 7

Special Considerations in Induction Heat Treatment

Use of induction heating for hardening and tempering of steel parts is now common in industry. As discussed previously, there are well-accepted methods for selection of generator frequency, power rating, coil design, and so forth, for a great many applications. Before applying induction techniques, however, special consideration should be given to the properties of the finished product. In previous chapters, we have touched upon several of these properties, such as hardness and strength. In the present chapter, other important characteristics shall be dealt with, including residual stresses, part distortion and quench cracking, scaling and decarburization, and temper brittleness.

DEVELOPMENT OF RESIDUAL STRESSES IN AXISYMMETRIC PARTS

One of the most important considerations in the design of a heat treatment process, be it an induction- or furnace-based one, is the development of residual stresses. The exact pattern of these stresses will depend on the heat treating temperatures employed, the depth of hardening, and the type of quench. Process conditions which give rise to compressive residual stresses on the surface of heat treated components are favorable. This type of stress delays the initiation of fatigue cracking in service, which typically occurs on the part surface under the action of cyclic tensile stresses. In addition, compressive residual stresses are known to delay the process of stress-corrosion cracking, a corrosion failure mode which is accelerated with increasing levels of surface tensile stress.

Residual Stresses in Subcritically Heat Treated Parts

The types of residual stress patterns that are commonly developed in axisymmetric parts such as turbine shafts and automotive axles when they are heated below the A_1 temperature can be rationalized by examining those patterns that are produced in simple round cylinders. These distributions have been described in detail by Johnson (Ref 73) and will only be summarized here. Referring to

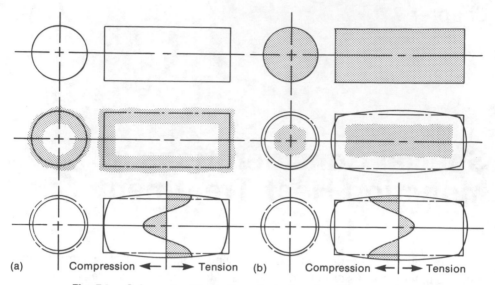

Fig. 7.1. Schematic illustration depicting development of residual stresses in steel bars (a) surface heated and (b) through-heated below the transformation temperature (Ref 73)

Fig. 7.1, consider the residual stress patterns developed in steel cylinders which are either (a) surface heated or (b) through-heated below the transformation temperature. In the surface-heated cylinder, a layer of hot metal is produced at the surface; this layer of metal is soft, and, therefore, the cold core has a tendency to "upset" or deform it to maintain continuity within the cylinder. When the workpiece is cooled, the surface, being hotter, tends to shrink more than the center. The center portion restrains this shrinking, putting the surface into a state of residual *tension* stress. Conversely, the surface, tending to shrink, induces *compressive* residual stresses in the center portion of the cylinder. The magnitude of the stresses depends on the depth of the surface-heated layer, the temperature to which it is heated, and the thermal and elastic properties of the material. In no case, however, can they exceed the yield strength of the material.

In the second case in Fig. 7.1, the entire cylinder is initially heated uniformly. During cooling, the temperature of the surface layers drops first, causing the metal here to become harder and tending to upset the central portion of the cylinder. When this central portion cools, it acts like the heated outer portion in the previous example. It puts the surface into *compression* and develops a *tensile* residual stress within itself, as shown schematically in the figure. For heating below the transformation temperatures, the magnitudes of these residual stresses in steels increase with the heating temperature (e.g., see Fig. 7.2), but these stresses must also remain lower than the yield strength of the alloy.

The examples in Fig. 7.1 pertain to what happens when the cylinder is cooled gradually. If a surface-heated cylinder is *quenched*, however, the residual stress

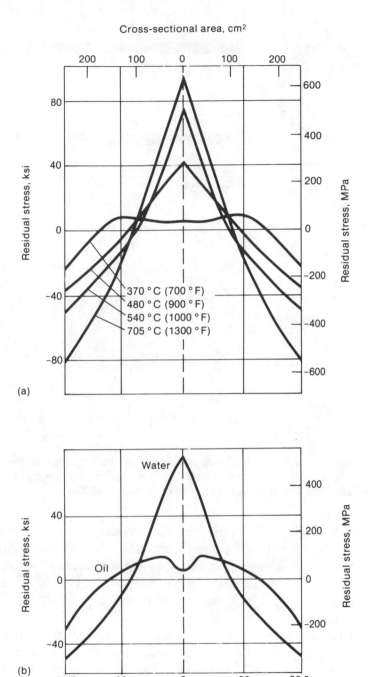

Fig. 7.2. **Longitudinal residual stress patterns developed in 17.8-cm (7.0-in.) diameter 1045 steel bars (a) heated to various temperatures and water quenched and (b) heated to 540 °C (1000 °F) and quenched in oil or water (Ref 74)**

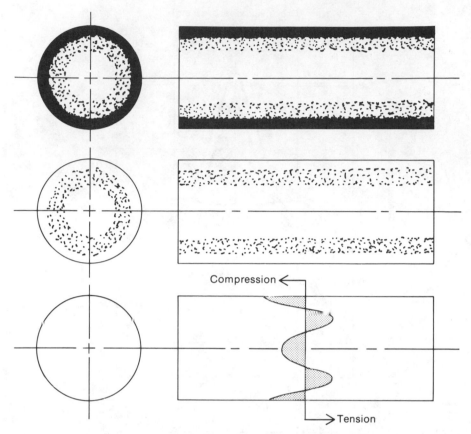

Fig. 7.3. Schematic illustration showing development of residual stresses in steel bars surface heated below the transformation temperature and water quenched (Ref 73)

pattern is modified (Fig. 7.3). In this case, the outermost layers of metal are cooled very rapidly. These layers upset those immediately beneath them. Thus, when these lower layers cool more gradually, they put the layers at the surface into compression. The difference between the residual stress patterns in the two surface-heated cases can be seen by comparing the stress distributions in Fig. 7.1(a) and 7.3.

The magnitudes of the residual stresses in a quenched cylinder initially through-heated are also greater than those in one slowly cooled, and the effect increases with the severity of the quench. As shown in Fig. 7.2, greater residual stresses are developed in water-quenched parts than in oil-quenched parts.

Residual Stresses in Austenitized and Hardened Parts

In the process of surface or through-hardening of steels, an additional factor must be considered in determining the magnitudes of the residual stresses — the

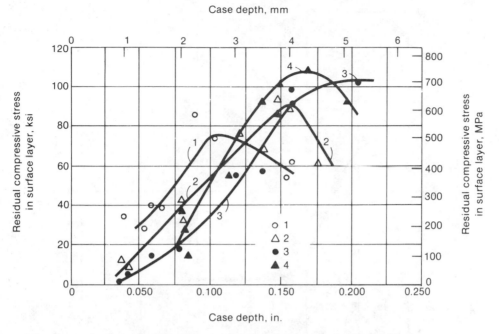

Case depth, mm

Steel compositions were as follows: curve 1, 0.44C-0.24Si-0.73Mn; curve 2, 0.12C-0.20Si-0.45Mn-1.3Cr-4.45Ni-0.85W; curve 3, 0.39C-0.26Si-0.65Mn-0.68Cr-1.58Ni-0.16Mo; curve 4, 0.38C-0.28Si-0.99Mn-1.33Cr-0.36Mo.

Fig. 7.4. Longitudinal residual stresses developed in the surface layers, as a function of case depth, in bars of four steels surface hardened by induction (Ref 75)

volume expansion associated with the formation of martensite. Because of this, a surface-hardened layer will have an increment of compressive residual stress in addition to, and often much greater than, the stress due solely to the quenching process. Another factor is the depth of the surface-hardened layer. As shown in Fig. 7.4, the magnitude of the compressive residual stress at the surface of an induction hardened bar tends to increase with increasing case depth. This increase is balanced by an increase in the tensile residual stresses below the surface. Also, not surprisingly, the depth to which the compressive residual stresses penetrate is usually about equal to the depth of the hardened layers (Fig. 7.5).

When parts are through-hardened, different residual stress distributions are developed. In this instance, the surface layers of the austenitized part transform first to martensite, leading to volume expansion of both the surface and the still-ductile inner core of austenite. When this inner core transforms subsequently, its expansion is prevented by the hard surface layers. The final result is, therefore, tensile residual stresses on the surface and compressive ones in the interior.

Because the residual stress patterns developed in through-heated parts are exactly opposite to each other when thermal-contraction effects versus trans-

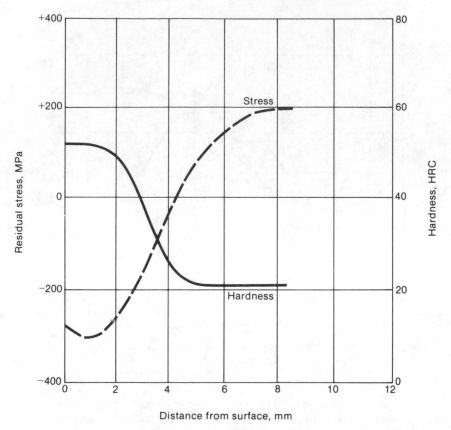

Fig. 7.5. Typical variation of hardness and residual stress with depth below surface for steel bars surface hardened by induction (Ref 76)

formation effects predominate, hardenability and section size can play an important role in the final residual stress pattern. This is illustrated by the data in Fig. 7.6 for a German steel similar in composition to an AISI 88xx series steel with moderate hardenability. Here, the cooling behavior and C-T diagrams are presented with measured residual stresses for three bar sizes. For the largest bar, martensite formation is avoided totally and the residual stress pattern is controlled largely by thermal-contraction effects. For the smallest-diameter bar, martensite forms throughout the section and tensile surface residual stresses and interior compressive residual stresses are formed. In the intermediate-size bar, in which martensite and intermediate-temperature transformation products are formed, a pattern between these two has been developed.

Other Residual Stress Considerations

Up to now, it has been implicit that the residual stresses which have been discussed have been longitudinal ones. Ordinarily, circumferential residual stresses

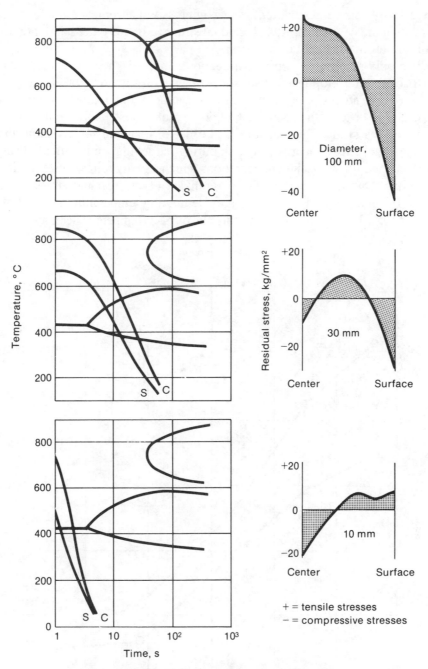

Chemical composition: 0.22% C, 0.65% Mn, 0.25% Si, 0.035% P, 0.035% S, 1.05% Cr, 0.45% Mo, ≤ 0.60% Ni.

Fig. 7.6. Continuous cooling behavior and residual stress patterns developed in bars of the German steel DIN 22CrMo44 in various diameters (Ref 77)

comparable in size and sign (tensile or compressive) accompany the longitudinal ones which are developed. Radial residual stresses, on the other hand, are usually small and are most likely to occur only in parts of large cross section.

When only a portion of a part is induction hardened, the adjacent regions may be expected to develop residual stresses also, which should be taken into consideration during process design. For instance, the circumferential compressive residual stress in a surface-hardened region would most likely induce a *tensile* residual stress in the region next to it in order to maintain stress equilibrium.

All residual stress patterns that arise during hardening are modified during the tempering and finish machining operations which usually follow austenitizing and quenching. During tempering, the magnitudes of the residual stresses are decreased by an amount that depends on the tempering temperature, as illustrated in Fig. 7.7 for a 1030 steel. These data are for furnace-tempered specimens. If these

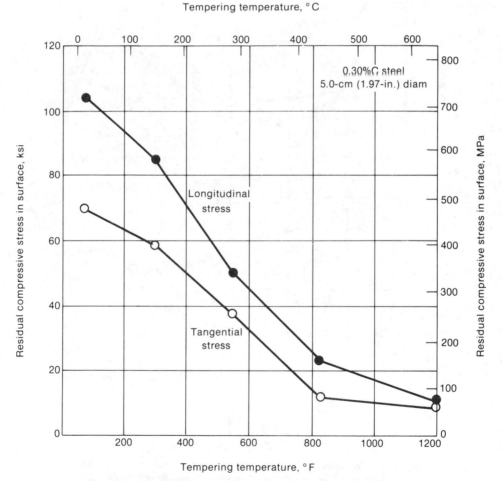

Fig. 7.7.   Effect of tempering temperature on longitudinal surface residual stresses in 0.30%C steel cylinders heated to 850 °C (1560 °F) and water quenched; tempering time, 1.5 h (Ref 75)

trends apply to short-time induction tempering treatments as well, however, the final residual stresses in induction-quenched-and-tempered steels can be expected to be lowered substantially during tempering. This is because of the higher temperatures mandated by the process.

Final grinding of hardened-and-tempered parts may also detrimentally affect the residual stress patterns developed during induction hardening. Surface compressive residual stresses are often reduced, or even changed to tensile ones, during such operations. Factors which contribute to this behavior include heat generation during the operation, producing effects similar to those of tempering, and plastic deformation of the surface layers. The latter effect is analogous to the effects of deformation during the development of surface tensile stresses, when only surface layers of a piece of metal are heated. An idea of the magnitude by which residual stresses can be changed can be obtained from the data in Fig. 7.8.

Provided that substantial surface compressive residual stresses are retained in the finished part, large increases in fatigue resistance can be realized. The increase in fatigue life in surface-hardened parts is a direct result of the compressive stresses and not the increase in hardness. Therefore, increases in compressive residual stress magnitude and depth of compressive stress penetration are beneficial. Both of these factors increase with case depth, as shown in Fig. 7.4 and 7.5. An example of the fatigue behavior associated with a bar hardened to different depths is given in Fig. 7.9, from which it may be observed that an increase in case depth

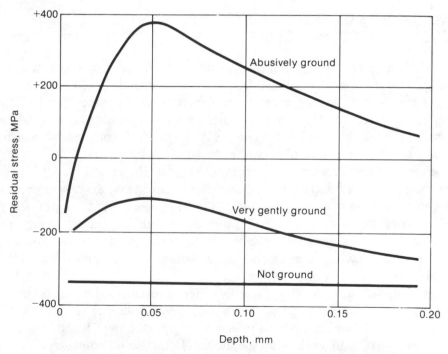

Fig. 7.8. Effect of grinding severity on surface residual stresses in induction quenched and tempered 1046 steel bars (Ref 76)

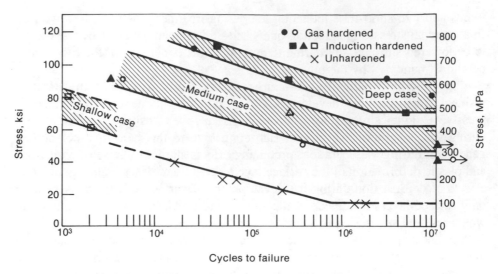

**Fig. 7.9. Fatigue S-N curves from reversed bending tests of un-
hardened and surface hardened steels of different case depths
(Ref 78)**

can increase the fatigue life (in terms of cycles to failure) by one or several orders
of magnitude.

DEVELOPMENT OF RESIDUAL STRESSES IN INDUCTION HARDENED GEARS

The surface hardening of gear teeth is another common application of induction
heating. A hard surface is beneficial from a wear standpoint. In addition, residual
stresses are also developed. For gears, the surface residual stresses may be either
compressive or tensile. The former are preferable in preventing bending-fatigue
failure of gear teeth in much the same way that surface compressive residual
stresses retard fatigue-crack initiation in shafts.

The residual stress pattern for a tractor drive gear, shown in Fig. 7.10, is the
most common type of distribution. As in axisymmetric parts, the development of
surface compressive residual stresses is due in large measure to the lower density
of the martensite which forms during quenching following austenitizing. The
magnitude of the compressive stresses may be a sizeable fraction of the yield
strength, thus allowing considerably higher stresses to be applied than could
ordinarily be done in fatigue loading.

Other work has suggested that markedly different residual stress patterns can be
produced in induction hardened gear teeth. In an investigation by several Japanese
workers (Ref 79), surface residual compressive stresses were found, but these
stresses *increased* with depth below the surface. In another investigation (Ref 80),
it was found that by controlling the input power density of the induction setup, the
magnitude of the surface residual stresses could be controlled. Lower power

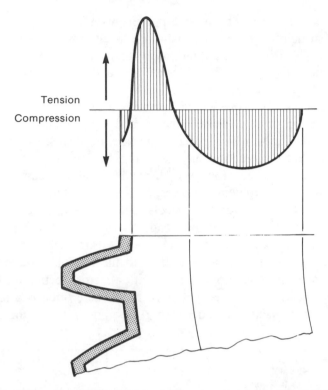

Fig. 7.10.   Qualitative residual stress pattern developed in induction surface hardened gear teeth (Ref 73)

densities, which tend to produce deeper cases, gave rise to larger compressive residual stresses. This effect is similar, therefore, to the effect of case depth on residual stress development in cylindrical parts. In the latter investigation, however, it was also discovered that very high power densities caused *tensile* residual stresses to develop. In this instance, the effect of martensite formation must have been outweighed by those of highly localized thermal expansion and plastic deformation. More work to document such trends is certainly warranted.

DISTORTION AND CRACKING DURING INDUCTION HEAT TREATMENT

Much as in furnace heat treating, distortion and cracking are important considerations in the design of induction heat treatment processes. Distortion arises during austenitizing or quenching. Distortion during austenitizing usually results from relief of residual stresses introduced during forging, machining, etc., or from nonuniform heating. When the part is only surface austenitized and hardened, the cool metal in the core of the workpiece minimizes distortion. Small amounts of distortion in induction surface hardened parts with shallow cases are often eliminated by means of a subsequent mechanical sizing (e.g., straightening) operation. Furthermore, the use of induction scanning, in which only a small portion of the workpiece is heated at any one time, is helpful in preventing problems of this type. Scanning is also helpful in keeping distortion levels low in through-hardening applications. In these instances, rotation of the part, provided that it is symmetrical, enhances the uniformity of heating and decreases the likelihood of nonuniformities in the final shape.

Distortion resulting from quenching is largely a function of the austenitizing temperature, the uniformity of the quench, and the quench medium. Higher austenitizing temperatures, which give rise to higher residual stresses, increase the amount of nonuniform contraction during cooling. Severe quenches such as water or brine, which also tend to produce high residual stresses, can lead to severe distortions as well. This problem can be especially troublesome when alloy steels are quenched in water. However, these steels usually have sufficient hardenability such that oil can often be employed instead.

In extreme cases, such distortions may lead to cracking. This cracking is intimately related to part design, as well as to the residual stresses which are developed. Components with large discontinuities in cross section are particularly difficult to heat treat, for this reason. In addition, there often is a limiting case depth beyond which cracking will occur (e.g., see Fig. 7.11); in these instances,

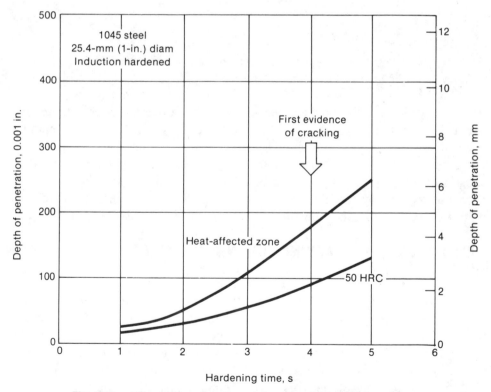

Fig. 7.11.   Effect of heating time on hardened depth and tendency toward quench cracking for machined and ground 1045 steel bars induction heated (450-kHz, 50-kW generator) and water quenched to a surface hardness of 62 HRC (Ref 50)

tensile stresses near the surface of the induction hardened part, which balance the compressive residual stresses generated, can be blamed for the cracking problem.

Steel composition also plays a role in the tendency toward cracking in induction hardening applications. This tendency increases as the carbon or manganese content is increased. This is not to say, however, that critical levels of either element can be specified, because other factors such as case depth (in surface hardening applications), part design, and quench medium are also important. The effect of carbon content on the tendency toward quench cracking is greatest in through-hardened parts and arises because of its influence on (1) the depression of the M_s temperature and (2) the hardness of the martensite (Fig. 7.12). As the carbon content is increased, the M_s temperature is lowered. Because the transformation occurs at lower temperatures, larger surface tensile residual stresses are produced. These residual stresses have a greater effect at lower temperatures, at which ductility is lower, than at higher temperatures. Moreover, the ductility of higher-carbon martensites at a given temperature is lower than that of lower-carbon martensites. Therefore, high tensile residual stresses and low ductility both contribute to quench cracking. However, inasmuch as hardenability increases with carbon content, quench-cracking problems often can be minimized by employing lower cooling rates (through the use of a milder quenchant) or by decreasing the difference between the austenitizing and quenching temperatures, both of which

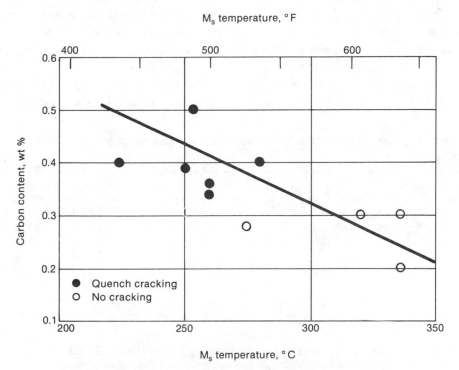

Fig. 7.12. Effect of carbon content on M_s temperature and occurrence of quench cracking for 1.3%Mn mild steel seamless tubing that was induction austenitized and water quenched (Ref 81)

tend to reduce residual stresses. Sometimes, however, higher-hardenability steels must be used to take advantage of these alternatives. Also, in some situations, quench-cracking problems can be alleviated by switching to lower-carbon or lower-manganese steels. It has also been found (Ref 82) that steels which are deoxidized* with aluminum rather than silicon exhibit a lower propensity toward quench cracking.

DIMENSIONAL CHANGES RESULTING FROM HEAT TREATMENT

Even if distortion and quench cracking do not pose problems, the dimensional changes that occur as a result of induction heat treatment should be evaluated, particularly if little or no machine finishing is to follow hardening and tempering of close-tolerance parts. Dimensional changes developed during hardening operations have two main sources — residual stresses and volume changes associated with phase transformation. For surface-hardening applications, the dimensional changes associated with these effects are usually minimal. On the other hand, through-hardening applications can produce non-negligible size changes. Those due to residual stresses (arising from thermal or phase-change effects) vary of course with material, part shape, and processing parameters. On the other hand, dimensional alterations resulting from phase changes alone can be estimated from the data in Table 7.1. Here it can be seen that volume change is greatest when a martensitic structure is produced in a part with a starting spheroidized microstructure. Less change can be expected after hardening of a part with an initial bainitic or fine pearlitic structure.

In contrast to the dimensional changes associated with hardening, those due to tempering generally involve volume decreases. This trend occurs when carbon is

Table 7.1.   Changes in volume that occur during various phase transformations (Ref 83)

Transformation	Change in volume, %
Spheroidized pearlite → austenite	$-4.64 + 2.21(\% \text{ C})$
Austenite → martensite	$4.64 - 0.53(\% \text{ C})$
Spheroidized pearlite → martensite	$1.68(\% \text{ C})$
Austenite → lower bainite	$4.64 - 1.43(\% \text{ C})$
Spheroidized pearlite → lower bainite	$0.78(\% \text{ C})$
Austenite → upper bainite	$4.64 - 2.21(\% \text{ C})$
Spheroidized pearlite → upper bainite	0

*Deoxidation is a term applied to the practice of adding elements such as aluminum or silicon to steels during melting and casting. These elements combine with oxygen in the steel and terminate the effervescent action caused by the formation of CO and CO_2 which results in the formation of "pipe" cavities in cast ingots. Deoxidized steels are also commonly referred to as "killed" steels because of the elimination of this effervescence.

precipitated from solution in martensite, leading to a higher-density micro-structure of ferrite and carbides. In steels in which austenite is retained following quenching, however, formation of bainite or pearlite from the austenite may produce volume increases during tempering. For these reasons, tempering may give rise to over-all dimensional decreases *or* increases.

SCALING/DECARBURIZATION

Because induction hardening and tempering are conducted in air (without pro-tective atmospheres), some consideration should be given to the magnitude of scaling and decarburization. Scaling can give rise to substantial material loss, an important economic consideration which is discussed in Chapter 10. Decar-burization during austenitizing can have a substantial effect on hardenability as well as on subsequent tempering response and, therefore, on final properties such as fatigue behavior of surface-hardened shafts.

For typical heating times, scaling and decarburization are negligible at tem-pering temperatures. On the other hand, they can be measurable at austenitizing temperatures.

At austenitizing temperatures, scaling (oxidation of iron and steels when heated in air) leads to the formation of the oxides FeO, Fe_2O_3, and Fe_3O_4. The first of these, FeO, or wustite, forms the bulk of the brittle oxide layer in the hardening regime. As with heat treating processes, time and temperature both play a role in determining the amount of material loss, as demonstrated by the data in Fig. 7.13 for scaling of 1020 steel when heated in air. It is apparent that the rate of scale formation (measured in terms of weight loss, Δm) is greatest early in the heating operation and then progressively decreases. The trend for this steel (and other steels) is often described by the parabolic law $(\Delta m)^2 = K_p t$, where Δm is ex-pressed in terms of weight per unit area, t is the exposure time, and K_p is the parabolic rate constant which is a function of material and temperature. For pure iron, the dependence of K_p on temperature has been found to be $K_p = 0.37 \exp(-33,000/R^*T)$ $(g^2/cm^4 \cdot s)$, when T is in kelvins (K) and R^* is the gas constant $(1.98 \text{ cal/g} \cdot K)$. The rate constants for steels are functions of the amounts of carbon and other alloying elements. In carbon steels, however, the rate of oxi-dation is often less than that for iron, as illustrated in Fig. 7.14.

An order-of-magnitude estimate of the scaling losses due to induction heating can be obtained from the rate constant for iron quoted above. As shown in Table 7.2, induction surface-hardening treatments (t ≈ 5 s) and induction through-hardening treatments (t ≈ 5 min) result in substantially less scale than furnace-based processes (t ≈ 1 h).

Further examination of the results in Fig. 7.14 reveals an unusual drop in scaling rate for the carbon steels at temperatures around 1050 °C (1920 °F). This behavior may be attributed to the competing effects of decarburization and scale formation. Decarburization involves the formation of CO and CO_2 from the carbon in the steel. It has been suggested that this reaction results in a decrease

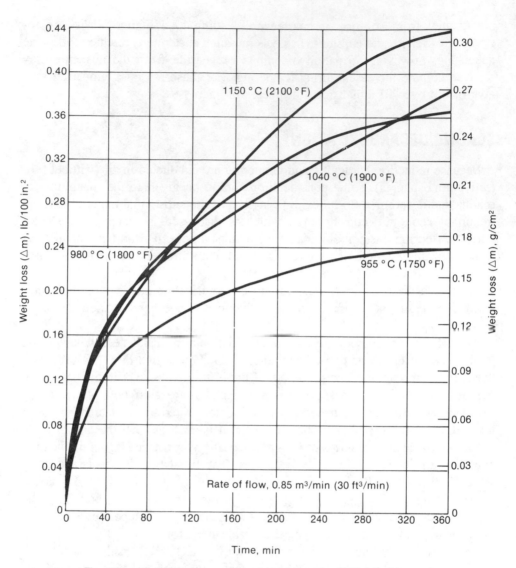

Fig. 7.13. Scale loss as a function of time for 1020 steel heated in air at various temperatures (Ref 84)

Table 7.2. Approximate depths of scaling resulting from heating of steel in air

	Depth of scaling(a) for temperature of:			
	955 °C (1750 °F)		1040 °C (1900 °F)	
Heating time	**cm**	**in.**	**cm**	**in.**
5 s	1.96×10^{-4}	7.7×10^{-5}	3.03×10^{-4}	1.19×10^{-4}
5 min..........	1.52×10^{-3}	5.99×10^{-4}	2.35×10^{-3}	9.24×10^{-4}
60 min.........	5.27×10^{-3}	2.07×10^{-3}	8.12×10^{-3}	3.19×10^{-3}

(a) Based on $K_p = 0.37 \exp(-16,667/T)$, where T is in kelvins (K).

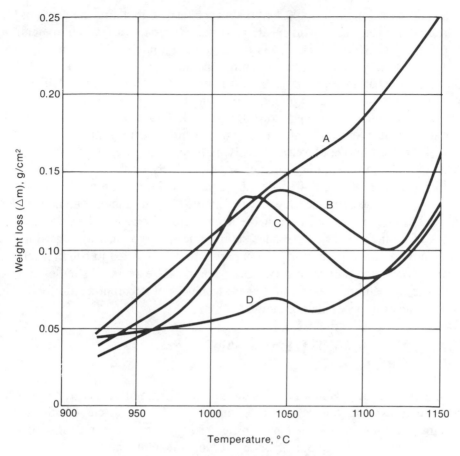

A – Armco iron; B – 0.15%C steel; C – 0.47%C steel;
D – 0.90%C steel.

Fig. 7.14. Effect of carbon content and temperature on scaling of carbon steels (0.6Mn-0.2Si-0.03S-0.02P) heated for 1 h (Ref 85)

in rate of diffusion of iron out of the base metal and of oxygen into it (Ref 85). This decrease gives rise to a thinner scale layer which is more highly oxidized (i.e., Fe_2O_3 and Fe_3O_4 are formed instead of FeO).

Decarburization during induction hardening depends strongly on time and temperature also. As stated by Birks and Jackson (Ref 86), decarburization tends to be minimal during heating in air below 910 °C (1670 °F). This is the temperature at which α transforms to γ for pure iron (ferrite), as seen in Fig. 2.2. During heat treatment below this temperature, the surface layers of a steel part are decarburized until they consist solely of α-ferrite. The diffusion of carbon through ferrite is relatively slow. Therefore, the surface ferrite layer in effect acts as a barrier to further decarburization.

For heat treatments in air above 910 °C, Birks and Jackson derived a solution of the diffusion equation to describe decarburization. Using the position at which the carbon concentration drops to 90% of that in the interior to define the depth of the decarburized layer, these workers derived the expression d (in centimeters) = $0.686 \sqrt{t} \exp(-8140/T)$. Here, t is the time in seconds, and T is the temperature in K. For a typical surface-hardening operation, t = 5 s, and T = 1223 K (950 °C); the decarburized layer is thus estimated to be 0.00197 cm (0.00078 in.). For a through-hardening application in which t = 5 min and T = 1173 K (900 °C), d is found to be equal to 0.0115 cm (0.0045 in.). In furnace through-hardening applications, in which times are an order of magnitude greater, the depth of the decarburized layer can easily be three or four times as large as this.

The above decarburization estimates are most likely on the high side, as indicated by measurements on 4037 steel by Birks and Jackson in which the actual thickness of the decarburized layer was found to be only about one-half of the predicted values. However, even if the decarburized zone is only several thousandths of an inch thick, its presence should not be forgotten with regard to the service environment, particularly if the part is to be subjected to bending fatigue or wear, for which a hard surface under compressive residual stresses due to martensite formation is beneficial. Severe instances of decarburization following austenitizing may necessitate grinding.

260 °C (500 °F) EMBRITTLEMENT AND TEMPER BRITTLENESS

In Chapter 2, the loss of toughness through tempering of quenched martensitic microstructures in certain temperature ranges was briefly discussed. Such embrittlement cannot be detected by tensile tests, but rather is manifested by a decrease in impact energy at ambient and cryogenic temperatures, or an increase in the ductile-to-brittle transition temperature (DBTT). The two most important forms of embrittlement are those which result from tempering at temperatures around 260 °C (500 °F), called 260 °C (500 °F) embrittlement, or at temperatures of 400 to 550 °C (750 to 1020 °F), resulting in an embrittlement known as temper brittleness.

The lower-temperature form of embrittlement has been associated most often with the formation of platelet carbides in conventional, long-time tempering treatments. The platelet carbides apparently form elongated and weak interfaces along which brittle cracks propagate easily. It has been found that when *short-time* induction heating is employed, the carbides formed during tempering are finer, more evenly distributed, and, most importantly, globular in shape. Therefore, one would not expect 260 °C (500 °F) embrittlement to occur under induction heating conditions. This hypothesis was verified by Nakashima and Libsch (Ref 87). The results of their study are summarized in Fig. 7.15, in which the DBTT's of 3140 steel specimens furnace tempered for 1 h at various temperatures or induction tempered (with zero hold time at temperature) were measured. For the same hardness levels obtained by tempering in the embrittling regime, the furnace tempered specimens had noticeably higher transition temperatures than the induc-

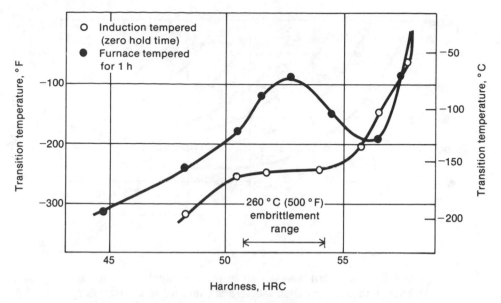

Induction tempered (t ≈ 0); Furnace tempered (t = 1 h).

Fig. 7.15. Comparison of impact transition temperature as a function of hardness level for 3140 steel tempered by induction and furnace methods (Ref 87)

tion tempered ones. Moreover, the induction tempered steel exhibited a continuous decrease of transition temperature with decreasing hardness (and, therefore, increasing tempering temperature). This trend is in line with the lack of an embrittling reaction in induction tempering heat treatments.

Libsch and his co-workers (Ref 88) also demonstrated the efficacy of short-time induction heating in eliminating temper brittleness. This embrittlement phenomenon is most often attributed to the diffusion of tramp elements, such as lead, bismuth, antimony, and tin (which are very difficult to eliminate totally from steel), to grain boundaries in martensitic microstructures. The short tempering times that are possible with induction prevent these diffusion processes from occurring. Results demonstrating the potential benefits of induction are given in Fig. 7.16. Here, the increase in the transition temperature relative to that for the unembrittled condition is shown for 1050 steel as a function of tempering temperature and time. It is obvious that treatments lasting approximately 10 min or less at any temperature in the embrittling range are not in the least detrimental. Such tempering times represent the upper limit of those normally used in induction tempering practice.

TEMPERATURE MEASUREMENT

The success of any induction heat treating operation depends a great deal on temperature measurement and control. In this section, methods of measuring

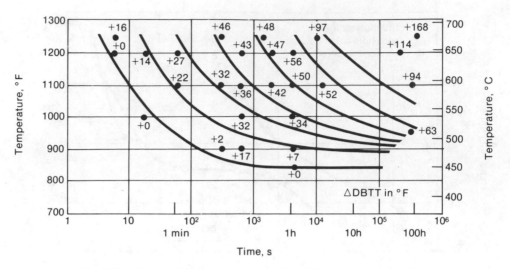

Fig. 7.16. Iso-embrittlement curves for 1050 steel showing increase in impact transition temperature relative to unembrittled condition (small numbers in plot) as a function of tempering temperature and time (Ref 88)

temperature are discussed. The control of temperature and automation of induction heating operations are described at the end of the chapter.

Methods of monitoring temperatures during induction heating have been reviewed extensively in the technical literature (Ref 89 to 92). The most common techniques make use of thermocouples and radiation detectors. Despite the widespread use of both methods, several major problems should be considered before a final selection is made. These include problems of poor workpiece surface condition, contact resistance, and response time for thermocouples. With radiation devices, emissivity variations are the major cause for concern.

Thermocouples

Although thermocouples are not suitable for all applications, they provide good accuracy, measurement capability over a very broad temperature range, ruggedness, reliability, and low cost. Thermoelectric thermometry is a mature technology whose principles have been known since the early 1800's. It is based on the well-known relationship between a difference in junction temperatures and the resulting voltage (emf). In practice, the reference junction is held at a constant known value by various means: e.g., an ice bath, a controlled-temperature furnace, or an electrical method of simulating a known temperature. The temperature of the heated junction is determined by measuring the voltage and referring to calibration tables for the particular thermocouple materials. Thermocouples are of two basic types: contact and noncontact (or proximity).

Contact Thermocouples. There are various types of contact thermocouple arrangements that permit accurate temperature measurements at fast response times.

The simplest and probably the most reliable technique involves direct attachment of a thermocouple to the part whose temperature must be determined. This procedure is impractical for measuring workpiece temperatures in most induction heating processes because of the presence of an induction coil (single-shot and scanning applications) or the fact that the part is moving through the coil (scanning methods of induction heating). For this reason, contact thermocouples are normally used only during initial trials for process setup.

The two most common types of contact themocouples are those in which the junction is welded to the specimen (referred to above) and those which are composed of two spring-loaded prods. In such an "open-prod" thermocouple (Fig. 7.17), the junction is made through the workpiece after contact is made. The prods can be made from any pair of thermocouple materials (e.g., chromel-alumel); the prods must be properly spaced to obtain the desired temperature/time response. Good electrical and thermal contact must also be established to produce accurate temperature measurements.

The open-prod thermocouple is most effective with metals which do not oxidize readily. However, by equipping the thermocouple assembly with a gas outlet port, it is possible to spray the ends of the prods and the contact area with a flux mixed with a combustible gas, thereby preventing or eliminating scale formation. The open-prod thermocouple is not capable of measuring temperatures of metals harder than the thermocouple elements themselves. These harder metals include steels and titanium- and nickel-base alloys. In these instances, noncontact thermocouples are appropriate.

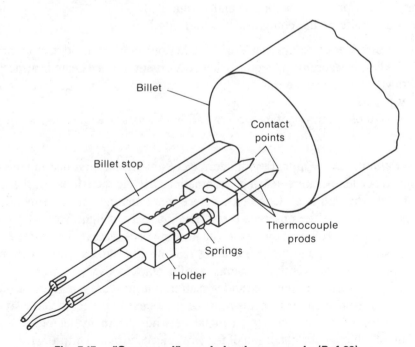

Fig. 7.17. "Open-prod" proximity thermocouple (Ref 89)

Noncontact (Proximity) Thermocouples.   To overcome the difficulties associated with scale formation, proximity thermocouples are often utilized. In this type of device, the thermocouple junction is attached to a disk of stainless steel or other high-temperature alloy. The disk (with the thermocouple attached to it) is welded on the inside and near the end of a stainless steel tube which acts as the temperature probe. By this means, the temperature-measuring device is isolated from the dirt and scale associated with the heating process. The distance between the disk and the workpiece surface is such that thermocouple temperatures do not unduly lag the actual surface temperatures. The disk is heated primarily by radiation from the hot surface; conduction through the entrapped gases and conduction through the tube also contribute to the total heat received by the disk.

Radiation Detectors

The other popular means of rapid temperature measurement in induction heating applications relies on the use of radiation detectors. These devices provide a noncontact method of measuring and controlling the temperatures of hot surfaces during heating.

In comparison with thermocouples, radiation detectors provide the following advantages:

- Because contact is not required, temperatures can be measured without interfering with the heating operation. Measurements can be made conveniently.
- There is no upper temperature limit for these devices.
- The response during measurement is very fast. The readings are accurate if the instrument is properly calibrated and maintained.
- The service life of the equipment is indefinite.

Insofar as disadvantages are concerned, radiation temperature detectors are expensive. Also, measurements require direct observation of the heated surface. The latter drawback can be overcome to a great extent by the use of modern fiber optics in many instances.

The two most common types of radiation devices are optical and infrared pyrometers.

Optical Pyrometers.   Optical pyrometers were among the first instruments used for noncontact temperature sensing. In practice, the operator looks at the incandescent body through a telescope-type device that contains a wire filament in the same optical plane as the observed body. The intensity of light from the filament is adjusted until the filament disappears against the background. The current to the filament is measured, and the temperature reading is obtained from a calibration of filament current versus the temperature of a black body.

Optical pyrometers are portable and versatile, but most are not highly accurate. The readings from such instruments must be corrected to reflect the emissivity of the workpiece. The emissivity of a metal depends on many factors, the most important of which is the condition of the workpiece surface. Unfortunately, these factors vary with temperature and heating conditions. In addition, smoke and water vapor in the air affect optical pyrometer readings. Because of such variables, use of fixed correction factors are not usually feasible. Nevertheless, at very high temperatures, typical of induction hardening applications, they are usually considered accurate enough for process control.

Infrared Pyrometers. Infrared pyrometers are rapidly replacing optical pyrometers for many applications. They are more accurate (0.5 to 1% of reading as opposed to ±2% of full scale for optical pyrometers), and they can be used as portable sensors or as part of a permanent, continuous temperature-monitoring system. In addition, the readout can be either analog or digital. Although there are many variations, the infrared pyrometer basically operates by comparing the radiation emitted by the hot target with that emitted by an internally controlled reference source. The output is proportional to the difference in radiation between the variable source and the fixed reference.

There are three types of infrared pyrometers:

- *Single-color or single-band pyrometers,* which measure infrared radiation from a fixed wavelength. These are the most versatile pyrometers because they are suited to many applications. Their accuracy may be affected by the presence of dust, atmospheric gases, and other characteristics of the measuring environment, and the emissivity of the target object must be taken into account.
- *Broad-band pyrometers,* which measure the total infrared radiation emitted by the target. The accuracy of such pyrometers is also adversely affected by the presence of dust, gases, and vapors. Similarly, errors in measurement can occur unless the emissivity of the target is taken into account.
- *Two-band or two-color pyrometers,* which measure the radiation emitted at two fixed and closely spaced wavelengths. The *ratio* of the two measurements is used to determine the temperature of the target. The adverse effects of the environment (dirt, gases, and vapors) are largely eliminated with this pyrometer. Moreover, because emissivity affects both measurements equally, its effect on accuracy is eliminated. Thus, oxides or scale on the workpiece introduce only minimal errors.

Of the three types of pyrometers, the two-color variety is surely the most accurate. As with the other types of pyrometers, it is readily adaptable for measuring temperatures in induction heating systems using a variety of optical systems for focusing and, as mentioned above, fiber optics.

Fiber-optic probes vary, in terms of both size and composition, from one application to another. Generally, they are less than 5 mm (0.2 in.) in diameter and can be fitted with lenses for sighting through gaps in the induction coil at very localized areas. Long-focal-length lenses are also frequently employed to provide adequate distance between the workpiece and the fiber optics and thus prevent overheating of the latter. Fiber-optic probes capable of withstanding temperatures of approximately 500 °C (930 °F) are available. At higher temperatures, probes may be equipped with air jets to keep them cool and clean of workpiece scale.

Other Temperature-Measuring Techniques

Thermocouples and pyrometers are the most widely used devices for measurement of temperature in industrial induction heating situations. Other techniques are currently under development and may find some use in the future. These include ultrasonic methods in which transducers are used to generate and detect sound waves in the workpiece. The velocity at which sound travels through a given material is a function of elastic modulus and density. Because these properties vary with temperature, the speed of sound through the material can be used to infer its temperature. Transducers under evaluation to determine their ability to generate the required acoustic waves include electromagnetic-acoustic transducers, electromechanical transducers, and high-intensity lasers.

In a related area, CAT (computer-aided tomography) is being investigated to determine its potential for visualizing internal temperature profiles of steel cylinders. To do this, ultrasonic data or measurements of the changes in resonant frequencies as a function of temperature are examined. This information is analyzed to construct graphic internal temperature profiles.

TEMPERATURE CONTROL

The devices described in the previous section are sometimes used not only to measure temperature but also to control it during induction heating operations which are designed specially to allow their utilization. In the simplest arrangement, the temperature-measuring device, be it a thermocouple or radiation detector, produces an output voltage, which is amplified and compared with a preset voltage, or "set point" (Fig. 7.18). The preset voltage corresponds to the desired temperature. The error signal is used to turn the induction power supply on or off, thereby controlling the temperature.

Modern control instruments are of two basic types: on/off and proportional (Ref 95). As the name implies, on/off types call for either full heating power or none at all. During heating, the power remains fully on until the set point is reached and then switches off until the temperature drops below the set point. The heat then switches fully on again. The primary disadvantage of this type of control action is the constant cycling of the temperature above and below the set point. However, it is useful in single-shot induction heating operations in which parts are heated to a certain temperature and then quenched or allowed to air cool, in which case the power is not switched on and off repetitively.

Induction scanning operations in which parts are fed continuously through the induction coil (and many single-shot processes also) usually rely on proportional controllers. These devices utilize three types of actions (proportional, rate, and reset) which serve to minimize temperature fluctuations. The proportional action is the control mode by which an output, or control, signal is generated which is proportional to the magnitude of the difference between the workpiece and set-point temperatures. The function of the proportional action alone would be similar

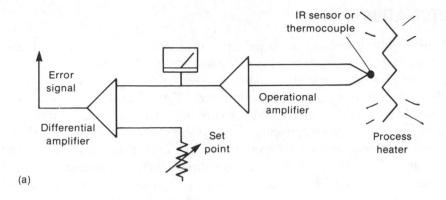

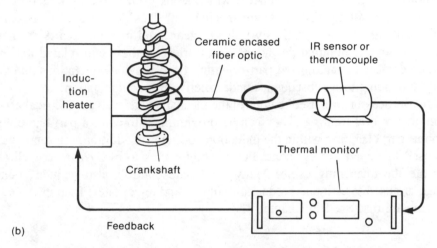

Fig. 7.18. **(a) Electrical and (b) system schematics for a simple temperature control system (Ref 93, 94)**

to that of the on/off controller except for a modulation added to the control action. In other words, as the temperature of the workpiece reached the set point, the control would decrease the power input to the coil to prevent overheating or underheating.

The proportional action alone, however, would provide no means of sensing the amount of heat (or power setting) required to maintain a given set-point temperature. Because of this, the temperature would tend to oscillate. In order to get the controller to equilibrate at the desired temperature, two other control actions — the rate and reset actions — are employed. The rate action takes into account how fast the actual workpiece temperature is changing in relation to the set point. A large rate of change in workpiece temperature would result in a large change in the control signal — larger than if just a proportional action were used. The reset action takes into account the time the actual workpiece temperature is away from the set point: the longer the time, the larger the correction.

AUTOMATION

Until relatively recently, most induction heat treating processes relied heavily on manual labor and/or relatively simple part-handling devices. In the past decade, a large number of completely automated systems have been developed and marketed by induction equipment manufacturers. Typical examples of these newer systems have been discussed by Miller (Ref 96). Besides the induction coil and part fixture, the major components of the system often include robots and microprocessors. Robots are used for material handling in the performance of such tasks as (1) unloading parts from previous operations (such as machining processes), (2) loading them into the induction fixture, and (3) moving them to a quenching, cleaning, or other work station such as an induction tempering area.

Another material-handling concept which lends itself to high-production automation is the in-line walking beam mechanism. This device and its associated equipment can also automatically unload, clean, and heat treat parts at rates in excess of several hundred per hour. Accessories for these and related machines include devices for automated part-orienting from previous operations and automatic unloading to a subsequent nondestructive inspection station.

Microprocessors are used to control the mechanical as well as electrical aspects of induction heat treating. They can be programmed to control part handling by a robot, part clamping within the induction coil, etc. In addition, microprocessors are used to impart closely controlled heating cycles to every part as well as to activate the quenching cycle. Many solid-state controls also include counters which are used to estimate production volume and reject rates. Examples of such systems are discussed in Chapter 8.

Chapter 8

Applications of Induction Heat Treatment

Since its introduction in the 1930's, induction heat treatment has been applied to a large variety of mass-produced, commercial products. The initial applications involved hardening of the surfaces of axisymmetric steel parts such as shafts. Subsequently, surface-hardening techniques were developed for other parts whose geometries were not so simple. Several typical surface-hardened parts are shown in Fig. 8.1. Most recently, induction hardening and tempering techniques have been developed for purposes of heat treating to large case depths and heat treating of entire cross sections. Types of parts to which induction is commonly applied include the following (Ref 97):

Surface-Hardening Applications
- *Transportation field:* crankshafts, camshafts, axle shafts, transmission shafts, splined shafts, universal joints, gears, valve seats, wheel spindles, and ball studs
- *Machine-tool field:* lathe beds, transmission gears, and shafts
- *Metalworking and hand-tool fields:* rolling-mill rolls, pliers, hammers, etc.

Through-Hardening Applications
- Oil-country tubular products
- Structural members
- Spring steel
- Chain links

In this chapter, applications and advantages of induction methods of heat treatment for some of the parts listed above will be discussed.

SURFACE-HARDENING APPLICATIONS

Crankshafts

Crankshafts for internal-combustion engines were probably the first parts to which induction hardening techniques were applied. Because the explosive forces

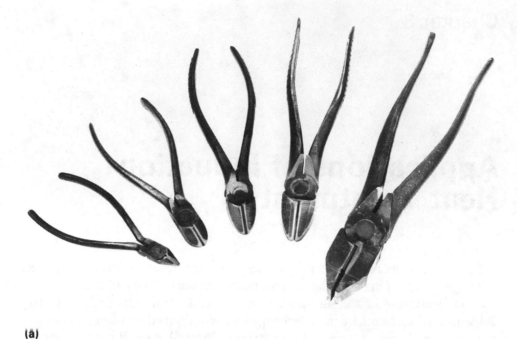

(a)

(b)

(a) Diagonal cutters and pliers surface hardened on cutting and gripping edges (extent of light-etching region indicates hardened zone). (b) Ball stud used in automotive suspension (section shows darker-etching hardened layer in ball portion of part). (c) Camshaft (section shows darker-etching, selectively hardened areas). (d) Equipment used in hardening the ball stud shown in (b); this includes the collets, coils, and fiber-optic temperature-sensing probes. Note that the quench-restraint tube has been removed from the right-hand fixture. [Photographs (a), (b), and (d) courtesy of S. Zinn, Ferrotherm, Inc., and American Induction Heating Corporation. Photograph (c) courtesy of Ajax Magnethermic Corporation.]

Fig. 8.1. Typical induction surface hardened parts

(c)

(d)

Fig. 8.1   (Continued)

of the engine must pass through the crankshaft, severe demands in terms of strength and wear resistance are placed on the steel used in manufacturing the crankshaft. These demands are ever-increasing with the rising horsepower ratings of engines used in automobiles, tractors, and other vehicles.

The most stringent demands are placed on the journal and bearing surfaces. Journals are the parts of the rotating shaft which turn within the bearings. Prior to the advent of induction heating, methods such as furnace hardening, flame hardening, and salt bath nitriding were used. However, each of these processes presented problems such as inadequate or nonuniform hardening and distortion.

Induction hardening overcomes many of these problems through rotation of the part during heat treating and selection of generator frequency and power to obtain adequate case depth and uniform hardness.* In one of the most common steels used for crankshafts, 1045, case hardnesses of about 55 HRC are readily obtained. Other advantages of the induction process for crankshafts include:

1. Only the portions which need to be hardened are heated, leaving the remainder of the crankshaft relatively soft for easy machining and balancing.
2. Induction hardening results in minimum distortion and scaling of the steel. The rapid heating associated with induction heat treating is advantageous in avoiding scaling in other applications as well.
3. Because induction heat treating processes are automated, an induction tempering operation immediately following the hardening treatment is readily feasible.
4. The properties of induction hardened crankshafts have been found to be superior to those of crankshafts produced by other techniques. These properties include strength and torsional and bending fatigue resistance. These improvements have enabled crankshafts to be reduced in size and weight.

Presently, crankshafts are being made from steel forgings as well as from iron castings. In the latter case, surface hardness levels of 50 HRC are easily obtainable after induction heating and air quenching. The resultant microstructure is a mixture of bainite and martensite, the pure martensite phase being avoided altogether. Such a dual microstructure minimizes the danger of crack formation at holes and eliminates the need for chamfering and polishing in these regions. The air quench allows residual heat left in the workpiece to minimize quench stresses and to autotemper the bainite which forms during cooling. After a prescribed period of time, the air quench is followed by a water quench during which the martensite phase is produced from the remaining austenite.

Axle Shafts

The axle shafts used in cars, trucks, and farm vehicles are, with few exceptions, surface hardened by induction. Although in some axles a portion of the hardened surface is used as a bearing, the primary purpose of induction hardening is to put the surface under a state of compressive residual stress. By this means, the bending and torsional fatigue life of an axle may be increased by as much as 200% over that for parts conventionally heat treated (Fig. 8.2). Induction hardened axles consist of a hard, high-strength outer case with good torsional strength and a tough, ductile core. Many axles also have a region in which the case depth is kept very

---

*Because of the rapid heating and cooling inherent in induction hardening operations, nonuniform expansion and contraction (and residual stresses) are important considerations in crankshaft hardening. Of particular concern is distortion of oil holes, which is prevented by use of metal plugs inserted into the holes prior to heat treating. Similarly, hardened depth and quenching technique must be carefully controlled during hardening of shafts with splines and keyways to avoid residual-stress-induced distortions.

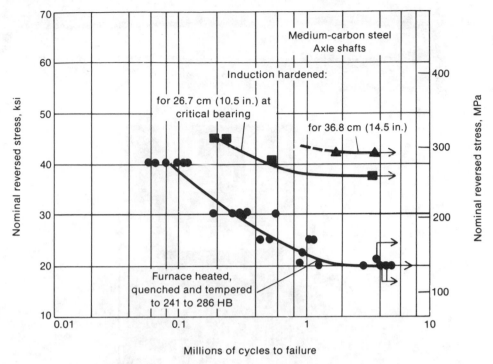

Shaft diameter, 7.0 cm (2.75 in.). Fillet radius, 0.16 cm (0.063 in.).

Fig. 8.2. Bending fatigue response of furnace hardened and induction hardened medium-carbon steel tractor axles (Ref 50)

shallow (Fig. 8.3) so that the part can be readily straightened following heat treatment. In addition to substantially improving strength, induction hardening is very cost-effective. This is because most shafts are made of inexpensive, unalloyed medium-carbon steels which are surface hardened to case depths of 2.5 to 8 mm (0.1 to 0.3 in.) depending on the cross-sectional size. As with crankshafts, typical hardness (after tempering) is around 50 HRC. Such hard, deep cases improve yield strength considerably as well.

Transmission Shafts

Modern transmission shafts — particularly those for cars with automatic transmissions — are required to have excellent bending and torsional strength besides surface hardness for wear resistance. Under well-controlled conditions, induction hardening processes are most able to satisfy these needs, as shown by the data in Fig. 8.4, which compares the fatigue resistance of through-hardened, case carburized, and surface induction hardened axles. The induction hardening methods employed are quite varied and include both single-shot and scanning (progressive hardening) techniques.

Nowadays, induction hardening of crankshafts, axles, and transmission shafts is becoming an increasingly automated process. Often, parts are induction hard-

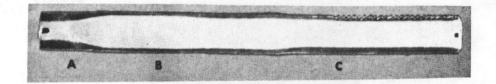

(b)

(a) Macrosection of axle showing region hardened for improved torsional strength (A) and wear resistance (C) as well as region with shallow case (B) to facilitate axle straightening. (b) Processing setup for induction surface hardening. [Photograph (a) from Ref 98. Photograph (b) courtesy of P. J. Miller, IPE Cheston.]

Fig. 8.3. Induction surface hardened 1037 steel axle

ened and tempered in-line. One such line, developed by Westinghouse Electric Corporation for heat treating of automotive parts, is depicted schematically in Fig. 8.5. It includes an automatic handling system, programmable controls, and fiber-optic sensors. Mechanically, parts are handled by a quadruple-head, skewed-

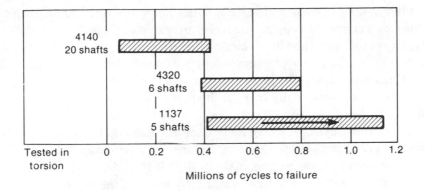

Steel	Surface hardness, HRC	Method of hardening
4140	36 to 42	Through-hardened
4320	40 to 46	Carburized to 1.0 to 1.3 mm (0.040 to 0.050 in.)
1137	42 to 48	Induction hardened to 3.0-mm (0.120-in.) min effective depth and 40 HRC

Arrow in lower bar (induction hardened shafts) indicates that one shaft had not failed after testing for the maximum number of cyles shown.

Fig. 8.4. Comparison of fatigue life of induction surface hardened transmission shafts with that of through-hardened and carburized shafts (Ref 50)

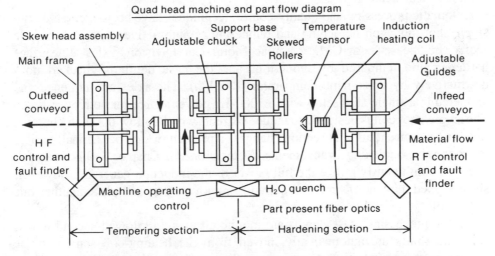

Fig. 8.5. Automated, quadruple-head, skewed-drive roller system used for in-line induction hardening and tempering of automotive parts (Ref 99)

drive roller system, abbreviated "QHD," after being delivered to the heat treatment area by a conveyor system. The roller drives, in conjunction with the chuck guides, impart rotational and linear motion to the incoming workpiece. Once a part enters the system, the fiber-optic sensor senses its position and initiates the heating cycle for austenitization. This sensor is also capable of determining if the operation is proceeding abnormally (e.g., if the part is being fed improperly) and can automatically shut down the system.

In the hardening cycle of the QHD system, the induction generator frequency is generally either in the radio frequency range (approximately 500,000 Hz) for shallow cases or in the range from 3000 to 10,000 Hz for deeper cases. In either instance, a temperature controller automatically senses if the part has been heated to a temperature too high or too low, in order to prevent an improperly austenitized piece from passing through the system. Assuming that the part has been heated properly, it then passes through a quench ring, which cools it to a temperature of 95 °C (200 °F) to form a martensitic case, prior to moving into the tempering part of the heat treatment line. Again, a fiber-optic system senses the presence of the part and begins the heating cycle using low-frequency current generally around 3000 Hz, since the desired tempering temperature is approximately 400 °C (750 °F) — a temperature at which the steel still has a large magnetic permeability. Once again, the part is automatically heated, quenched, and moved from the heat treatment station, this time onto a conveyor which takes it to the machining area for grinding.

The control system of this line is designed to allow decisionmaking by programmable controls. Thus, all aspects of the heat treating process and mechanical operations are preprogrammed and may be changed easily to accommodate different part sizes and heat treating parameters. With such a process, users have been able to increase production rates more than threefold over those obtainable with conventional heat treating lines.

Automatic processing and quality-control techniques have been integrated with single-shot induction hardening and tempering of automotive parts such as axle shafts. In one setup at GM's Cadillac Motor Car Division, axles are surface hardened and tempered by induction, and the case depth is checked nondestructively by a magnetic comparator (Ref 100). The axles are hardened and tempered in a single-shot setup with a 3000-Hz power source for both operations. The single-shot coil is designed to induce currents which follow the shaft contours rather than concentrating at section changes, a problem which occasionally arises in progressive hardening arrangements. As mentioned in Chapter 5, this technique does not continuously utilize the full power of the induction generator, however, since a fixed amount of time is required to remove a part and place another one in the coil.

During processing, case depths are monitored using a kilowatt-second meter, and axle shafts are automatically moved from the heating-coil station to the quenching station in both the hardening and tempering processes. Following heat treatment, parts are transported to a straightening station. Off-line is a magnetic comparator for determining case depth. The comparator consists of a testing coil and a low-frequency power source (e.g., for axle shafts, a frequency of 1 Hz is used). In much the same way that the inductor and workpiece interact during induction heating, the current in the coil of the magnetic comparator is affected by the electrical and magnetic properties of a part placed within it. Since these properties are affected by the microstructure of a steel part, the comparator can be calibrated to determine the depth of a martensite case. In this way, the quality of induction hardened components can quickly undergo a 100% inspection, which is very desirable for parts serving critical applications (such as axles).

Gears

Reliability and high dimensional accuracy (to ensure good fit) are among the requirements for gears such as those used in transmissions for farm equipment and related applications. Keeping distortion as low as possible during heat treatment is very important. Thus, induction is probably the best process for such parts. Among the other advantages of induction heat treatment of gears are the following:

- Gear teeth and roots can be selectively hardened.
- Heating is rapid with minimum effect on adjacent areas.
- Uniform hardening of all contact areas results in high wear resistance. The improvement of wear resistance often permits substitution of inexpensive steels, such as 1045 or 1335, for highly alloyed steels.

When using induction, however, extreme care is needed in positioning the gear in relation to the coil, particularly in setups in which all gear teeth are heated and hardened at once. In these instances, the coil goes entirely around the gear, and a quench ring concentric to it is used (Fig. 8.6a). A typical hardening pattern for this kind of arrangement is shown in Fig. 8.6(b). In such single-shot setups, a two-stage process is often preferable, however. In the first step, a relatively low frequency is used for heating of the root diameter of the gear and for partial heating of the flank areas between the roots and tooth tips. Then, the tooth tips themselves are heated with a much higher radio frequency. As with surface hardening of shafts, gears are usually rotated during processing to effect uniformity of heating and hardening. Part transfer between stations in this and similar processes is carried out by specialized systems or robots. In this way, a uniform or contoured hardness pattern which follows the outline of the gear is obtained. This hardness pattern improves not only the wear resistance of the teeth, but their bending strength as well. Modifications of single-shot techniques may be employed for preferential hardening of only certain portions of the teeth, such as tooth tips or flank regions, depending on specific applications. Unfortunately, as the gear becomes larger, the capacity of the induction generator needed for surface hardening increases dramatically, as shown in Table 8.1 for various gear geometries.

Another common technique for surface hardening of gears is the so-called tooth-by-tooth technique. As the name implies, each tooth is individually heated and quenched. By this means, induction generators of modest capacity can be used

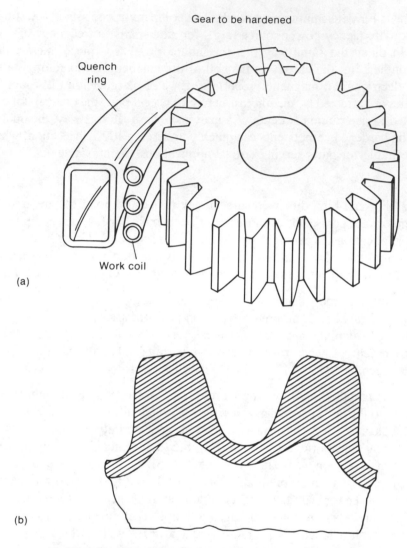

(a) Solenoid coil and concentric quench ring around gear to be hardened. (b) Schematic of case-hardness pattern obtained with such an arrangement.

Fig. 8.6. Induction hardening of gear teeth by a single-shot technique (Ref 48)

for large gears which otherwise would require large coils and large amounts of power. A typical inductor for such a process consists of a copper coil that is slightly larger than the gear tooth (Fig. 8.7a). Another coil design for this process is illustrated in Fig. 8.7(b). With this design, two adjacent tooth flanks are heated at any one time. To one side of the coil, there are two copper end plates and a laminated iron core. This is needed to overcome fringing effects during heating of thick gears in which the coil scans the tooth roots in the thickness direction. The

Table 8.1. Power requirements for induction hardening of gear teeth (Ref 101)

Tooth	Diametral pitch	Approximate length of tooth profile		Surface area per tooth(a)		Power required per tooth(b), kW	Total power required(c), kW
		cm	in.	cm^2	in.^{2}		
A	3	5.1	2.0	12.9	2.0	20	800
B	4	3.8	1.5	9.7	1.5	15	600
C	5	3.3	1.3	8.4	1.3	13	520
D	6	2.5	1.0	6.5	1.0	10	400
E	7	2.3	0.9	5.8	0.9	9	360
F	8	1.9	0.75	4.8	0.75	7.5	300

(a) For a face width of 2.54 cm (1 in.). (b) At a power density of 1.55 kW/cm^2 (10 kW/in.^{2}). (c) For a gear having 40 teeth.

actual shape of the coil depends to a large extent on the hardness pattern desired. If only flank and root hardening are desired, a coil whose outer corners are chamfered is employed, as shown in Fig. 8.8(a). Without the chamfers, the tooth tips would be heated as well, often in a very nonuniform manner. Alternatively, auxiliary water sprays, which are used for quenching, may be adjusted to cool the tooth tips (Fig. 8.8b), or the coil can be shortened.

Valve Seats

Valve seats in automobiles are yet another application of surface hardening by induction. Prior to the advent of catalytic converters and the need to use unleaded gasoline, wear resistance of valve seats was afforded by deposits of lead oxide. These deposits acted as a lubricant between the seat and valve. Without the lead oxide from gasoline, other means of preventing premature valve wear (usually within 10,000 miles) were required. In order to avoid the expense of hardened inserts, an induction heat treating method by which all the seats in a single engine head can be processed at one time was developed. This is done with a specially designed machine in which the surfaces are heated rapidly and self-quenched to produce a case depth of 1.8 to 2 mm (0.06 to 0.08 in.) and a hardness of 50 to 55 HRC. Figure 8.9 shows the improvement over untreated parts that such processing affords. The durability of the induction hardened seats is even superior to that of conventional seats in engines which use leaded fuel.

Railroad Rails

Surface hardening of railroad rails is one of the more recent applications of induction heat treatment. The heads (top portions) of rails wear rapidly in curved sections where high-tonnage freight-car traffic is common. The abrasive action of the wheels combined with high stresses can result in very short rail life, sometimes as little as one year or less. With the move toward heavier cars and increased speeds, these kinds of problems are becoming more severe.

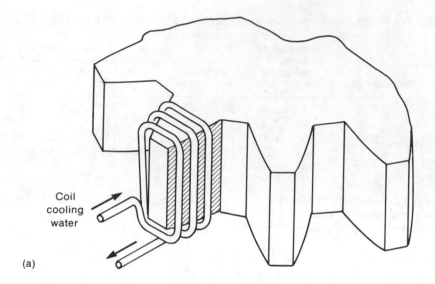

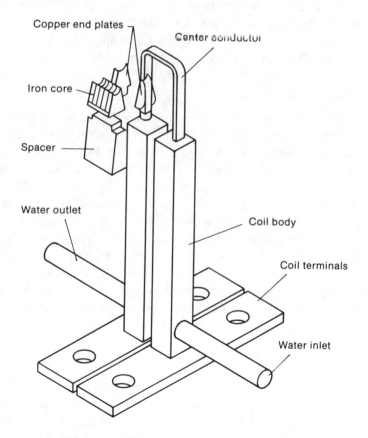

(a) Setup using a conventional coil. (b) Setup in which two adjacent tooth flanks are heated and hardened at one time.

Fig. 8.7. Arrangements for tooth-by-tooth hardening of gears (Ref 48, 102)

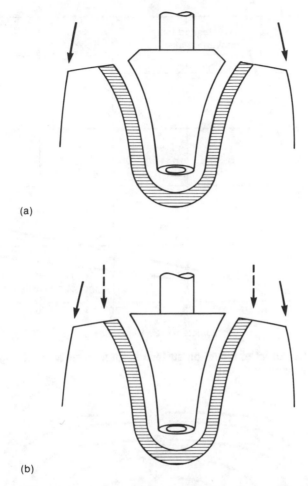

(a) Inductor design with chamfered corners. (b) Inductor design without chamfers in which tooth-tip hardening is prevented by adjusting water sprays. In both sketches, arrows indicate direction of preferred water spray: solid arrows in (a) and dotted arrows in (b).

Fig. 8.8. Inductor designs for tooth-by-tooth hardening of gear flanks (Ref 102)

Conventional railroad rails are manufactured from 1080 steel by hot (shape) rolling using a preheat temperature of 1290 °C (2350 °F). Following rolling, they are controlled cooled, and a finished product of only moderate hardness (250 HB ≈ 24.5 HRC) results. In the induction process, only the head of the rail is hardened since this is where failure takes place because of wear or deformation during service. A relatively thick case whose hardness decreases with depth (Fig. 8.10) is achieved by using a relatively low-frequency (approximately 1000-Hz) power source. In the actual process (Ref 65), rails are prebent (elastically) before heat treatment to offset distortions caused by heating and to eliminate the need for final straightening operations. The rails are then fed continuously

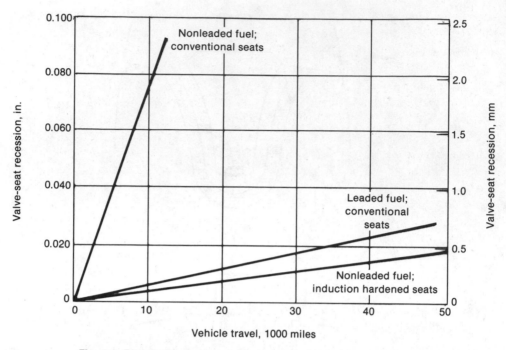

Fig. 8.9. Effect of induction surface hardening on wear of engine valve seats (Ref 103)

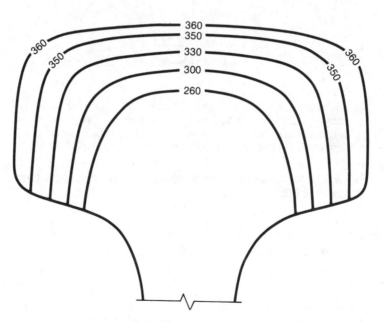

Fig. 8.10. Brinell hardness pattern in induction surface hardened railroad rail (Ref 65)

through a U-shape inductor and their surfaces heated to 1065 °C (1950 °F). Following heating, the surface is air quenched to 425 °C (800 °F), producing a bainitic microstructure. Residual heat left in the interior of the rail brings the surface layers back to a temperature of 595 °C (1100 °F), thereby bringing about autotempering. Finally, a controlled cold-water quench is applied to cool the rail to room temperature and to ensure straightness. Rails produced thereby have been found to last from 2½ to 8 times as long as conventionally manufactured rails.

Research by Russian investigators (Ref 104) has demonstrated that induction surface hardening in combination with autotempering leads to properties almost identical, and at times superior, to those obtained by induction hardening and furnace tempering. In this work, rails were heated to a surface temperature of 950 to 1000 °C (1740 to 1830 °F) using a frequency of 2500 Hz. The rails were then quenched (using a combination of air and water) to about 400 °C (750 °F), and autotempering was allowed to take place at 400 to 450 °C (750 to 840 °F) for 45 s prior to final quenching to room temperature. Some of the rails were straightened and tested; others were given a 2-h furnace tempering treatment at 460 °C (860 °F) prior to straightening and testing. Both batches exhibited a bainitic case and a pearlitic core. More importantly, the mechanical properties were almost identical. Figure 8.11 shows the equivalence of hardness and tensile properties. Such behavior is not surprising in view of the logarithmic time dependence of tempered hardness, as described in Chapter 6. In addition, the Vickers hardness distribution shown in Fig. 8.11(a), when converted to Brinell hardness, is almost identical to that in Fig. 8.10. Other measurements revealed that the fatigue resistance of the two lots was equal, and that the fatigue endurance limit was 30 to 40% higher than that of conventionally processed rails. The toughness of the induction hardened rails was also better than that of conventional rails, particularly for rails which were straightened and tested *without* additional furnace tempering. From the above, it may be concluded that induction surface hardening of railroad rails is a very beneficial process.

Rolling-Mill Rolls

Induction hardening of rolling-mill rolls is analogous to induction hardening of rails in that relatively deep cases are produced. During service, roll life is limited by abrasive wear. As the diameter is reduced by wear, adjustments are made to bring the rolls closer together in order to maintain a given rolling reduction. These adjustments are sufficient until the rolls have worn approximately 38 mm (1.5 in.); once this amount of wear is exeeded, the rolls must be replaced. The objective of induction heat treatment is, therefore, to produce a hardened case approximately 19 to 38 mm (0.75 to 1.5 in.) deep. This is done employing a low-frequency power supply. The 60-Hz equipment developed by Ajax Magnethermic for Bethlehem Steel Corporation typifies that ordinarily selected (Ref 105). This equipment is used for hardening by either a single-shot or progressive (scanning) technique.

In the progressive hardening method, the roll, hanging vertically, is lowered into the induction coil, in which its surface temperature is gradually raised to

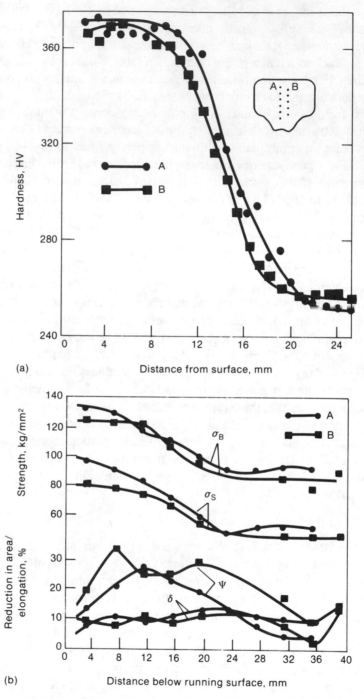

(a)

(b)

(a) Vickers hardness distributions across induction hardened rails either autotempered (A) or autotempered and subsequently furnace tempered (B). (b) Mechanical properties of the two lots of rails: σ_s = yield strength; σ_B = ultimate tensile strength; δ = elongation; ψ = reduction in area.

Fig. 8.11. Properties of induction hardened railroad rails (Ref 104)

Depth below surface, mm

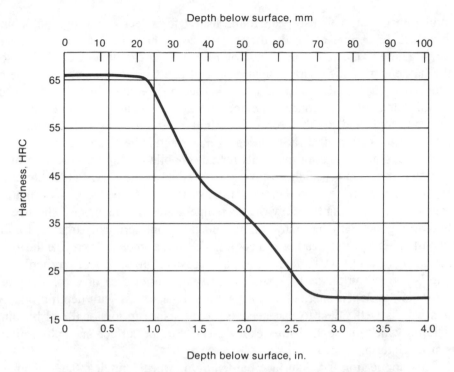

Depth below surface, in.

Fig. 8.12. Hardness pattern developed in rolling-mill rolls induction surface hardened using a 60-Hz generator (Ref 105)

955 °C (1750 °F). By controlling the power input and feed rate, a temperature profile is developed such that the temperature ranges from 900 °C (1650 °F) at 38 mm (1.5 in.) below the surface to less than 260 °C (500 °F) at 51 mm (2 in.) below the surface. Following heating, the roll is quenched using water precooled to 5 °C (40 °F). Because roll steels usually contain 0.8 to 0.9% C and substantial amounts of nickel, chromium, molybdenum, and vanadium, they have high hardenability and develop high hardness to the entire depth to which the steel was austenitized. A typical hardness profile is shown in Fig. 8.12. Here, the drop in hardness beyond about 25 mm (1 in.) can be attributed to heat losses due to conduction, which could have resulted in the formation of pearlite or bainite prior to quenching, at which time the remaining austenite would have transformed to martensite.

Miscellaneous Applications

There are many other applications of induction surface hardening. These include uses in the ordnance, hand-tool, and automotive fields.

In the ordnance area, induction heating has been used for both surface hardening and through-hardening of armor-piercing projectiles. The induction process allows a very uniform bainitic microstructure to be obtained. During World War II, it was found that batch furnace heat treatment could not produce as uniform and high-quality a product at such a low cost as could the induction method.

Induction heating has also been found useful for selective through-hardening and surface hardening of heads for tools such as hammers, axes, picks, and sledges. These tools are usually made of 1078 steel. Lead baths were once used in the hardening of such parts, but use of lead baths has diminished because of health and safely regulations. Induction hardening, because of its high speed, versatility, and ability to concentrate heat in selective areas (an aspect important in the hand-tool field), is therefore increasingly being used as a replacement.

Another application of surface hardening by induction involves the manufacture of high-strength steel spring materials for automobile coil springs, leaf springs, and torsion bars. One specialized process described in the U.S. Patent literature (Ref 106) consists of repeated heating of the surface above the A_3 transformation temperature and cooling to below the A_1 temperature. By this means, the steel is heated throughout with the central grains being only intercritically treated, for the most part. This repeated heating and cooling also has a tendency to refine the grain structure significantly. After quenching, a microstructure of fine-grain, hard martensite is formed on the surface, and a somewhat softer core of a coarser grain structure is developed internally. The variation of hardness with depth produced in this way is well suited to that needed for resistance to torsional and bending fatigue, in which the applied stresses are greatest at the surface and substantially less at the interior.

A final application is the surface hardening of wheel spindles. These components are the parts of the axle around which the wheels turn. As such, they have bearing surfaces which are subjected to large amounts of abrasion. Increased resistance to wear is thus afforded by selective surface hardening. Surface hardening also imparts improved bending fatigue resistance.

THROUGH-HARDENING APPLICATIONS

Although not as common as surface hardening and tempering, through-hardening and tempering via induction methods have been found to be practical for a number of applications such as piping, structural members, saw blades, and garden tools. Some of these applications shall be described next.

Through-Hardening and Tempering of Pipe-Mill Products

Probably the largest application of induction through-hardening (and tempering) involves piping or tubular goods used for oil wells and gas pipelines, for example. For these uses, relatively low-frequency induction generators are selected so that the reference depth is of the same order of magnitude as the wall thickness of the workpiece. Since the workpiece is hollow, there is no problem of loss of electrical efficiency arising from eddy-current cancellation such as the losses which occur at the centers of solid bars. In fact, to a point, the efficiency of induction heating of tubular products increases as the wall thickness decreases, because the resistance of the material increases with decreasing wall thickness and becomes much larger than that of the coil. However, if the thickness is very small, the current developed in the workpiece goes down and relatively little I^2R heat is generated.

Pipe-mill products fall into two major categories: electric resistance welded (ERW) and seamless. ERW pipe is made from steel strip which is formed and welded. After welding, the weld may be annealed to avoid cracking during shipment or subsequent operations, which may include reduction to obtain a smaller diameter or different wall thickness. In any case, ERW products tend to have a very uniform wall thickness, which is an important consideration in induction heating and heat treatment. Nonuniformities in wall thickness usually lead to temperature nonuniformities; thicker regions are heated to lower temperatures than thinner ones during induction heat treatment. In contrast to ERW pipe, seamless piping tends to have a much less uniform wall thickness. It is manufactured by piercing and extruding a heated, round-cornered square billet. To maintain quality, seamless piping for oil-country applications is typically required to have a wall thickness variation of no more than 12.5%. Because of its uniformity, ERW pipe is the preferred choice for induction heat treatment.

In a typical installation for heat treatment of piping, processing is carried out in a continuous line in which the steel is austenitized, quenched, tempered, and finally cooled to room temperature at successive stations. A typical arrangement is that employed by Lone Star Steel (Ref 107), depicted in Fig. 8.13. In this system, each pipe is loaded onto an entry table and fed onto the conveyor as soon as the heat treatment of the pipe preceding it is completed. As each pipe passes through the austenitizing station (consisting of five coils), it is rotated on skewed

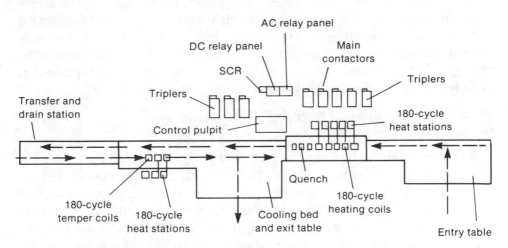

Pipe enters from the right, is austenitized, quenched, drained, and tempered. Following tempering, the pipe is transferred to cooling beds for air cooling.

Fig. 8.13. Schematic diagram of equipment used for in-line induction through-hardening and tempering of pipe-mill products (Ref 107)

rollers to ensure temperature uniformity. Also, because only a small portion of the pipe is heated at one time, distortion is readily controlled. Using 180-Hz current, pipes are heated uniformly through the thickness to approximately 900 °C (1650 °F). A suitable power density is chosen for this purpose as well. With a maximum generator capacity of 4500 kW, piping up to 406 mm (16 in.) in OD can be handled by the austenitizing unit. For this largest diameter, the 152-cm (5-ft) length of the heating section results in a maximum power density of roughly 0.23 kW/cm^2 (1.5 kW/in.^{2}), assuming 100% efficiency, when the total power of the generator is utilized.

After austenitizing, the pipe enters the quench ring several feet down the line. After the water has been drained off, the pipe moves to the tempering station which is powered by generators with a total capacity of 2700 kW and also at a frequency of 180 Hz. The power capacity for this operation is lower than that for austenitizing because the workpiece is heated to temperatures of only about 540 to 650 °C (1000 to 1200 °F). Following tempering, the pipe continues along the conveyor to cooling beds and is rotated during the entire cooling cycle to ensure straightness and lack of ovality.

A similar induction hardening and tempering line is in place at Tubulars Un limited, Inc. (Ref 70). This line also processes pipe-mill products for the oil-drilling industry which range from 60 to 270 mm (2.4 to 10.6 in.) in OD and from 4.3 to 38 mm (0.17 to 1.5 in.) in wall thickness. Austenitizing is accomplished using generators of 300 Hz (for heating to temperatures around the Curie point) and 500 Hz (for heating to temperatures from the Curie point to the austenitizing temperature). Tempering is accomplished with generators of the lower frequency. As in the Lone Star line, temperatures are continuously monitored and, if necessary, corrected. In this way, products with a high degree of uniformity through the wall thickness and which meet stringent American Petroleum Institute standards are manufactured. In these operations, electrical efficiencies approaching 95% enable production efficiencies of around 75%, which are substantially greater than those realized in batch-type furnace operations.

Through-Hardening and Tempering of Structural Members and Bar Stock

A process similar to hardening and tempering of pipe-mill products is used to heat treat structural members of uniform section thickness, as described in the patent literature (Ref 108). In these cases, the structural member is passed through a series of induction preheating and heating stages for austenitizing, and then is quenched while being restrained by a set of rolls which prevent distortion. The various coils for austenitizing are connected to generators of frequencies ranging from 180 Hz (preheating) to 10,000 Hz (final heating stages) in processing of steel shapes 6 to 13 mm (0.25 to 0.5 in.) in section thickness. After quenching, the structural shape is tempered in-line using induction heating frequencies of 180 to 3000 Hz and is prevented from distorting during subsequent cooling by another set of restraining rolls.

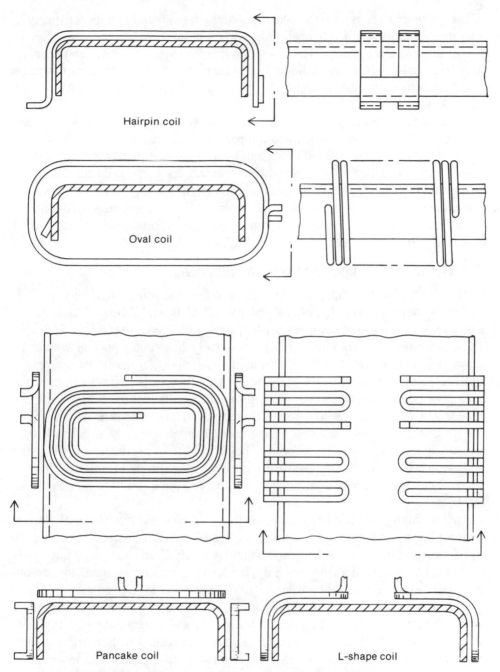

Fig. 8.14. Various coil designs used for induction through-hardening and tempering of structural members fabricated from steel strip and plate (Ref 108)

The above process is used to make high-strength structural members from 1025 steel strip in the form of U-channels, T's, and I-sections. For U-channels, a variety of coil designs are possible. These include hairpin, oval, pancake, and L-shape coils (Fig. 8.14), all of which induce eddy currents whose paths lie in the plane of the structural member and which follow directions similar to those of the currents in the coil.

In a related application, Nelson (Ref 109) demonstrated that induction heating could be used to through-harden and temper bar stock of various grades of alloy steels. Ranging in size from 20 to 90 mm (0.8 to 3.5 in.), bars were heat treated using 1- to 3-kHz generators, the smaller frequencies being used for the larger diameters. In order to ensure temperature uniformity through the thickness, low power densities and heating times on the order of minutes were employed with the result that very uniform properties and microstructures, often better than those of conventionally heat treated bars, were obtained.

Miscellaneous Through-Hardening Applications

Other induction hardening and tempering applications are often very specialized, requiring special coil designs and control of heating. One of these is the selective hardening and tempering of circular-saw-blade segments (Ref 110). These segments are required to have a relatively hard outer portion where the cutting teeth are located and a softer, tougher inner portion where the segment is attached to the periphery of the main body of the saw. It is desirable to have the hard outer portion extend back as far as possible in order that the cutting teeth can be repeatedly "dressed" as they wear. To accomplish this, the entire blade segment is through-hardened by induction. Using special coil designs, the segment is then selectively tempered, the "back" portion being softened a great deal more than the portion in which the cutting teeth have been milled. Through proper controls, a relatively abrupt transition zone between the two regions and a large hardened area can be achieved.

Another high-production application is the hardening and tempering of garden-trimmer blades and snow-plow blades. Typically made from steels such as 1070 and 1080, these blades are hardened to about 60 HRC and tempered back only slightly. The presence of a uniformly hardened zone allows them to be continuously resharpened as they wear, much like saw-blade segments.

One of the more interesting applications of induction through-hardening involves the manufacture of coil springs. For this product, steel rod stock, typically about 1.27 cm (0.50 in.) in diameter, is induction heated to hot working temperatures, put through a coiling machine, and finally quenched to martensite. The ability to perform the heating, forming, and hardening in a continuous line makes the use of induction particularly attractive in this instance, not only from a processing standpoint but also with regard to energy economy, since the coil springs do not have to be reheated for heat treatment.

Chapter 9

Other Heat Treating Processes Using Induction

In the previous chapters, we have focused on application of induction heating to hardening and tempering of low- and medium-alloy steel products. This can be considered the area of greatest use insofar as heat treating processes are concerned. Nevertheless, induction has also been employed for other kinds of heat treatments, including hardening of cast irons, annealing of aluminum and steel, and heat treatment of stainless steels and tool steels. Each of these applications will be discussed in turn.

INDUCTION HARDENING OF CAST IRONS

Cast irons (e.g., iron-carbon alloys which contain more than 2 wt % C), like steels, can be hardened and tempered through control of the decomposition of austenite into pearlite, bainite, or martensite, and the subsequent heat treatment of the transformation products. Unlike steels, however, cast irons cannot be processed to produce a uniform microstructure, inasmuch as the carbide phase (or the graphite constituent common to many alloys) is not totally soluble in the austenite phase at any temperature. This can be seen in Fig. 2.2. It is the ability to retain this second phase which makes these iron alloys almost impossible to forge and thus usually suitable only for casting. On the other hand, the presence of large graphite flakes, for instance, leads to a product which is readily machined.

Despite the inability to obtain a uniform austenite phase during the hardening operation, induction is still readily applied to many cast iron parts, such as cast crankshafts, to obtain hardened wear-resistant surfaces (55 to 60 HRC) and to avoid distortion. The choice of frequency and power density is made using guidelines similar to those for steels and is based on the size of the part and the case depth needed. Shallow cases require high frequencies and power densities, and deep cases require lower frequencies and power densities.

The starting microstructure is also an important consideration in induction hardening of cast irons. It is important that the carbon in the alloy will readily dissolve in the iron at the austenitizing temperature because of the short times inherent in induction processes. For example, the combined carbon in pearlite is readily soluble, and high percentages of pearlite (usually around 50 to 70% in gray iron and nodular iron) in the matrix are desirable; for this reason, cast irons are often normalized prior to hardening to obtain a pearlitic structure and to improve austenitizing response. Sufficient pearlite to give a combined carbon content of 0.40 to 0.50% is usually adequate to provide the hardnesses needed in cast iron parts. The combined carbon behaves essentially the same as carbon in steel of similar carbon content. On the other hand, the graphitic carbon in the form of large flakes (gray cast iron) or large nodules (nodular cast iron, also often known as ductile iron) remains essentially unchanged during heat treatment. As another example, in nodular cast irons which have been previously quenched and tempered, the carbon in the matrix is in the form of many fine nodules. When this microstructure is induction heated to hardening temperatures, the carbon dissolves and diffuses rapidly, enabling quick austenitization and excellent hardening response.

The principal difference between hardening of steels and hardening of cast irons lies in the need to control temperatures more closely for the latter materials. For steels, the austenitizing temperature can be varied over a fairly wide range without measurably affecting the hardening characteristics, provided that long soak times and austenite grain growth are avoided. In contrast, temperature control is very important for cast irons because the carbon content of the austenite phase for a given alloy increases with temperature. It will also increase with the percentage of pearlite in the starting microstructure. (For this reason, the hardening temperature is generally decreased with increasing pearlite content.) As the carbon content increases, the temperatures at which martensite begins to form (the M_s temperature) and at which the transformation is completed (the M_f temperature) are depressed. With sufficiently high carbon content, a substantial amount of soft austenite can be retained on quenching to room temperature. Cryogenic treatment can be employed in order to eliminate this problem. Alternatively, a tempering treatment at about 300 °C (570 °F) can also be used to transform the retained austenite to bainite, thereby increasing the hardness of the cast iron as well.

TRANSVERSE FLUX INDUCTION HEAT TREATMENT

Heating of strip products for purposes of heat treatment offers one of the largest potential markets for the induction process. Steel and aluminum are the materials most often fabricated into strip and then heat treated to produce properties needed for further fabrication into consumer articles in the automotive, appliance, and can industries. In conventional practice, the heat treatment is carried out by placing coils of the strip to be treated into large batch furnaces in which they are heated (and cooled) rather slowly, typical cycles lasting many hours and even days because of the large thermal inertia of the coils. An alternative process consists of continuous heat treatment in which strip is gradually fed through the heating unit.

Induction heating is rather well suited for such a process in view of its ability to heat small areas to high temperatures at high rates.

As might be expected, induction heating of thin strip requires careful coil design, selection of frequency and power, and process controls. Coils which surround the strip are incapable of developing sufficient skin effect at typical generator frequencies. The most efficient design is a pancake coil with which eddy currents whose paths lie in the plane of the sheet are set up. In this case, the lines of magnetic flux are perpendicular to the current paths (and the sheet) — unlike those for a coil which surrounds the sheet, in which case the flux lines are parallel to the sheet surface. These behaviors are illustrated in Fig. 9.1. Because of the flux pattern for the pancake coil, such a heating process has come to be known by the name "transverse flux induction heating."

In transverse flux induction heating of strip, uniformity of temperature is achieved by moving the strip through the inductor at a constant speed and by inducing current patterns such that every part of the strip receives an equal amount of energy. Current patterns and fringing problems are controlled to a certain extent by wrapping the induction coils around laminated iron cores or pole pieces. The laminations discourage the induction heating of the iron cores themselves. Moreover, the iron cores minimize the amount of current and number of coil turns needed to set up a given magnetic field and induce a given amount of eddy current in the workpiece. Computer programs have also been used to simulate the development of the electromagnetic fields and to estimate the effects of process variables such as line speed, generator frequency, coil design, and workpiece properties on temperatures achieved and temperature uniformity. By these means, processing can be controlled so as to avoid, to a certain extent, the overheating of strip edges. This was a major problem in early transverse flux lines. It resulted from eddy currents which traverse the sheet and go along the strip edges to form complete current paths.

As in other induction heating applications, over-all selection of generator frequency depends largely on the sheet thickness and the resistivity and relative magnetic permeability of the workpiece. Figure 9.2 gives frequencies recommended by Waggott and his co-workers (Ref 111) for transverse flux induction heating of aluminum and steel strip. For both materials, the frequency increases with decreasing thickness. Also, it can be seen that the recommended frequencies for aluminum are substantially lower than those for steel. This is because the resistivity of aluminum is much smaller.

Waggott *et al* also discussed one of the most successful transverse flux induction heating lines in existence. This line is used to anneal aluminum strip to be used in products such as aluminum cans and auto-body panels. The strip comes to the line in the cold rolled condition, in which it is hard but very uniform in thickness. Although some aluminum alloys are hardenable through heat treating, those for can and automotive applications typically are not, and induction heating is used to soften the cold rolled product to enable it to be formed further. To compensate for short heating times, the temperatures needed to carry out the

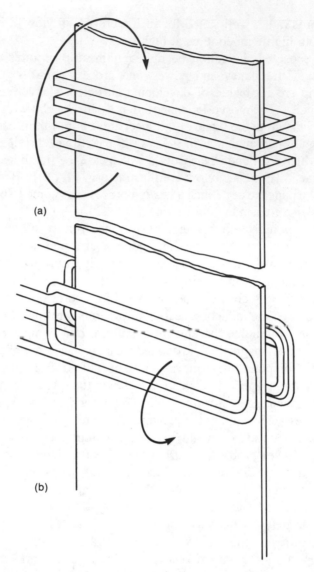

Arrows indicate lines of magnetic flux, which are (a) longitudinal
and (b) transverse (i.e., perpendicular to the sheet thickness).

**Fig. 9.1. Heating of strip metal by (a) solenoid coil and (b) pan-
cake coil (Ref 111)**

annealing operation using induction are somewhat higher than those used for
long-time batch annealing of aluminum.

A schematic illustration of the heat treating line used by Alcan Plate Ltd. (in
Great Britain) is shown in Fig. 9.3. The front end of the strip is "stitched" onto
the tail end of the preceding coil. In turn, the strip is then induction heated, passed
through a short soak zone, and water quenched. The water is removed using an
air knife, after which any distortion due to quenching is eliminated (by a tension

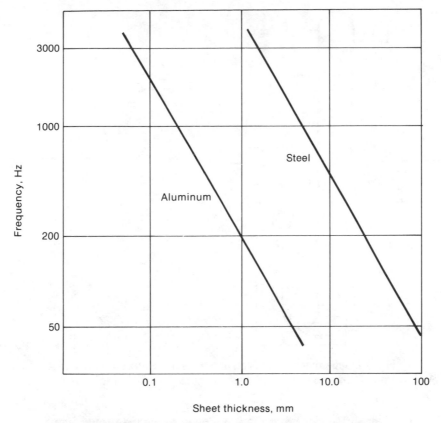

Fig. 9.2. Relationship between required generator frequency and sheet thickness for transverse flux induction heating of aluminum and ferritic steel alloys (Ref 111)

leveller) prior to coiling of the strip or further processing down the line. A controller automatically adjusts the induction generator output power to account for the speed at which the strip is passed through the inductor and the strip thickness. Correct operating conditions are also ensured by temperature measurements across the sheet width following heating. With this system, energy efficiency of approximately 75%, or more than twice that usually obtainable in batch furnaces, is realized.

The material produced in a unit such as that described above has properties as good as, or better than, those of material produced by batch annealing processes. Table 9.1 demonstrates this trend for two non-heat-treatable aluminum alloys, 3103 and 3105. For both alloys, the transverse flux induction heated (TFIH) material exhibited higher yield and tensile strengths and about equivalent ductility. Further, property measurements across the width and along the length of the strip showed them to be very uniform and always within 2% of those in the alloy specifications. It can also be noted from the data in Table 9.1 that TFIH yields a product with considerably finer grain size than batch-annealed material. This

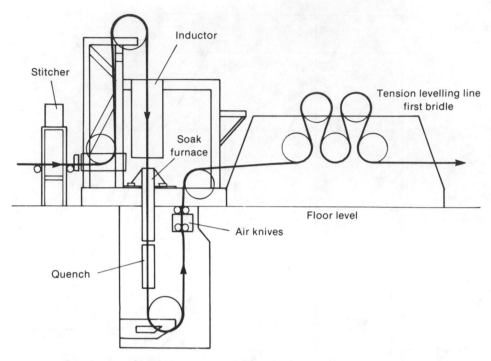

Fig. 9.3. Schematic illustration of a transverse flux induction heat treating line used to process aluminum alloy strip (Ref 111)

Table 9.1.   Comparison of properties achieved by conventional batch annealing and transverse flux induction annealing (Ref 111)

Alloy(a) and treatment(b)	Ultimate tensile strength		0.2% proof stress		Elongation, %	Grain size, μm
	MPa	ksi	MPa	ksi		
3103, batch annealed	114	16.5	44	6.4	33	53
3103, TFIH annealed	116	16.8	62	9.0	34	11
3105, batch annealed	121	17.5	38	5.5	25	41
3105, TFIH annealed	129	18.7	62	9.0	24	15

(a) Compositions (in wt %): 3103, Al-1.2Mn; 3105, Al-0.55Mn-0.50Mg. (b) TFIH = transverse flux induction heated.

characteristic is a result of the rapid heating and short soak time involved in TFIH. With such fine grain sizes, poor surface quality ("orange peel" effect), which may occur during cold forming of large-grain materials, is avoided.

The TFIH process has also been used to produce *partially* annealed and hardened aluminum alloy strip. Uniform partially annealed strip, difficult to manufacture using batch methods, is an attractive product because it offers a good blend of strength and ductility. It has also been found that many heat treatable aluminum alloys, which are given a high-temperature "solution annealing" treatment and a subsequent lower-temperature precipitation hardening treatment, can be satisfactorily processed without extended hold time at temperature. Careful control of composition and prior thermomechanical history are required, however, to ensure rapid solution annealing, a process which is quicker in alloys with a fine pre-existing dispersion of precipitates.

RECRYSTALLIZATION ANNEALING OF STEEL

Although the TFIH process has yet to be applied to continuous annealing of steel sheet, there is reason to believe that it would be just as successful for this material as it is for processing of aluminum alloys. In the case of steel, the largest application would be for low-carbon steel. Typically containing 0.05 to 0.10 wt % C, these steels are produced in large coils and used extensively in the automotive and appliance industries. At present, the majority of low-carbon steel for these uses is made in the form of cold rolled sheet which is furnace annealed after rolling in order to recrystallize the grain structure. Typically, the sheet has been cold rolled 50 to 80%. Small reductions on the order of 10 to 20% can lead to very undesirable grain growth during annealing. The furnace (or batch) annealing process for coils of low-carbon steel requires a long heat-up period (on the order of one day) because of the large mass. Annealing *per se* is done near the lower critical temperature, i.e., around 690 to 730 °C (1275 to 1350 °F) for times on the order of one day. During this soak, recrystallization and some grain growth occur. In addition, carbides, which have been formed during cooling of steel that was hot rolled prior to cold rolling, may redissolve. However, slow cooling of the coils after soaking promotes reprecipitation of these carbides. As annealed, the low-carbon steel is relatively soft and has a grain size of ASTM 6 to 8. Because the annealed steel has been slow cooled from the annealing temperature and because all carbon has come from solution, one of the major age-hardening problems in low-carbon steels — quench age hardening (or, for brevity, "quench aging") — is avoided. Quench aging leads to gradual hardening and decreasing ductility with time at room temperature due to carbide precipitation. With regard to subsequent forming of the steel sheet, such changes in properties are undesirable.

In contrast to furnace annealing, induction heat treatment of low-carbon steel for purposes of recrystallization usually consists of fast heating to temperature, little or no soaking at the maximum temperature, and cooling to room temperature. Some studies have shown that soak times between 1 and 20 s are needed for complete recrystallization (Fig. 9.4). Others have shown that no soak time is necessary provided that a high enough temperature is reached at the end of the heating cycle. In these instances, recrystallization starts at some lower temperature and progresses as the temperature is increased. The temperatures at which recrystallization begins and is completed for a 1010 steel cold rolled to an 80% reduction in thickness and heated at various rates by induction are shown in Fig. 9.5. Note that as the heating rate is increased, the recrystallization temperatures are also increased.

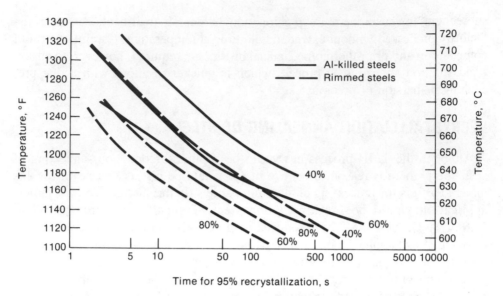

Kinetics were determined by salt bath annealing followed by water quenching.

Fig. 9.4. Time-temperature relationships for recrystallization of aluminum-killed and rimmed low-carbon steels cold rolled to the indicated reductions in thickness prior to annealing (Ref 112)

Rapid recrystallization heat treatments tend to produce a finer-grain product than furnace annealing. However, unless strip cooling following heating is controlled very carefully, the potential for undesirable effects due to mild to pronounced quench aging increases. The primary means of avoiding quench aging is slow cooling of the steel through temperatures at which carbide precipitation is most rapid. Usually, cooling for 10 to 15 s at a temperature in the range of 510 to 425 °C (950 to 800 °F) is sufficient for low-carbon steels. There are also reports (Ref 113) that, by rapid induction heating to 650 °C (1200 °F) and subsequent water quenching, cold rolled steel can be recrystallized without putting the carbides into solution. Thereby, a soft, ductile steel with minimal tendency toward quench aging is obtained. Such findings need to be verified, but further work is certainly warranted in order to realize the full potential of induction in continuous annealing of low-carbon steel sheet.

STRESS RELIEF/NORMALIZING OF PIPE WELDS

Stress relief and annealing of welds in electric resistance welded piping is another process for which induction heating finds great application. The weld area and the heat-affected zone are the regions treated in these instances. At these locations, nonhomogeneous microstructures characteristic of melted and solidified metal, and high residual stresses due to quenching following welding, are found. These conditions lead to poor ductility and the likelihood of brittle fracture during

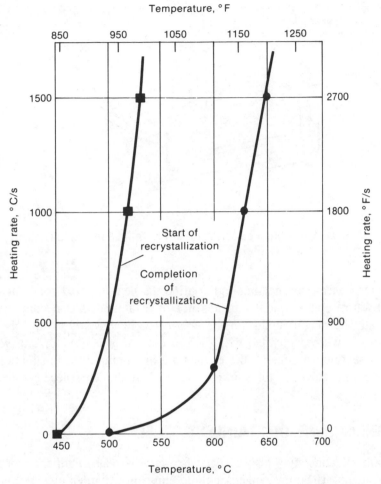

Sheet thickness, 0.05 cm (0.020 in.). Sheet was heated using an 800-Hz generator and a pancake coil.

Fig. 9.5. Effect of induction heating rate on temperatures for initiation and completion of recrystallization of low-carbon steel sheet cold reduced 80% prior to heat treatment (Ref 113)

service. To overcome these problems, the weld and heat-affected areas are heated to temperatures above the A_3 temperature (to produce an austenitic microstructure) and allowed to air cool. This "normalization" treatment eliminates residual stresses and results in a pearlitic microstructure of increased ductility which blends into the microstructure on either side of the weld. Pipes may then be hardened and tempered or put directly into service.

The inductor design usually selected for stress relief and normalizing is known as a split-return coil (Fig. 9.6). The physical size of the coil is determined by the size of the pipe and the area to be heat treated. Welds produced by low-frequency currents provide deep heating and lead to large heat-affected zones, whereas high

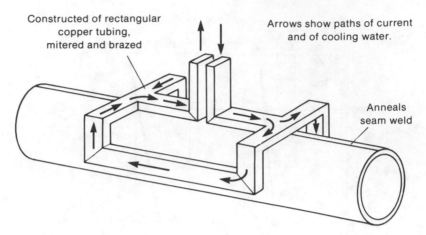

Constructed of rectangular
copper tubing,
mitered and brazed

Arrows show paths of current
and of cooling water.

Anneals
seam weld

Fig. 9.6. Split-return coil used for stress relief and normalizing of pipe seam welds (Ref 43)

frequencies concentrate heating and confine the heat-affected zone. In addition, the split-return coil frequently makes use of a laminated iron core in order to concentrate the magnetic flux in the area to be heated and to increase heating efficiency. It will be recalled that an iron core is also used in transverse flux induction heating. In general, these cores are often employed in situations where the coil does not surround the workpiece and where the workpiece geometry is not too complex.

HEAT TREATMENT OF STAINLESS STEELS

Induction heating may also be used for various heat treatment processes for stainless steels. These treatments include annealing, hardening, and tempering; the appropriate ones for different stainless steel grades are listed in Table 9.2.

The austenitic (300 series) and ferritic (e.g., 430, 446) stainless steels are not hardenable through heat treatment. However, induction heating is often used for annealing of these alloys to restore ductility after cold working and to enable further shaping. In these processes, temperatures 25 to 55 °C (50 to 100 °F) higher than those of normal furnace treatments are used to compensate for the very short heating and soak times.

Martensitic stainless steels (e.g., grades 410, 416, 420, 440A, 440B, and 440C: see Table 9.2) are similar to other AISI quenched-and-tempered steels in that they are magnetic and are hardened by austenitizing and tempering. Hardening temperatures must be selected carefully when using induction. Because chromium carbides go into solution rather slowly, these temperatures must be 55 to 95 °C (100 to 175 °F) higher than those of furnace treatments (typically done around 995 to 1025 °C, or 1825 to 1875 °F) to allow for the short heating times and to minimize excessive grain growth. Furthermore, induction tempering of martensitic stainless steels is usually performed at the high ends of the tempering

Table 9.2. Induction heat treatments for stainless steels (Ref 114)

AISI designation	Annealing Temperature, °C (°F)	Coolant	Hardness, HB	Hardening Temperature, °C (°F)	Coolant	Hardness, HRC	Tempering Temperature, °C (°F)	Coolant	Hardness
302...	1065 to 1150 (1950 to 2100)	Water	150	Nonhardenable			Nonhardenable		
303...	1065 to 1150 (1950 to 2100)	Water	150	Nonhardenable			Nonhardenable		
304...	1065 to 1150 (1950 to 2100)	Water	150	Nonhardenable			Nonhardenable		
309...	1095 to 1150 (2000 to 2100)	Water	160	Nonhardenable			Nonhardenable		
316...	1065 to 1150 (1950 to 2100)	Water	150	Nonhardenable			Nonhardenable		
321...	1010 to 1095 (1850 to 2000)	Water	160	Nonhardenable			Nonhardenable		
347...	1065 to 1150 (1950 to 2100)	Water	160	Nonhardenable			Nonhardenable		
410...	Generally furnace annealed			1010 to 1065 (1850 to 1950)	Oil or air	~43	205 to 650 (400 to 1200)	Air	97 HRB to 41 HRC
416...	Generally furnace annealed			1010 to 1065 (1850 to 1950)	Oil or air	~43	To 650 (To 1200)	Air	...
420...	Generally furnace annealed			1040 to 1065 (1900 to 1950)	Oil or air	~54	175 to 510 (350 to 950)	Air	48 to 52 HRC
440A...	Generally furnace annealed			1040 to 1120 (1900 to 2050)	Oil or air	56 to 57	175 to 510 (350 to 950)	Air	50 to 57 HRC
440B...	Generally furnace annealed			1040 to 1120 (1900 to 2050)	Oil or air	58 to 59	175 to 510 (350 to 950)	Air	54 to 59 HRC
440C...	Generally furnace annealed			1040 to 1120 (1900 to 2050)	Oil or air	59 to 60	175 to 510 (350 to 950)	Air	55 to 60 HRC
430...	790 to 845 (1450 to 1550)	Air	150 to 180	Generally not hardened			Generally not hardened		
446...	790 to 845 (1450 to 1550)	Air	150 to 180	Generally not hardened			Generally not hardened		

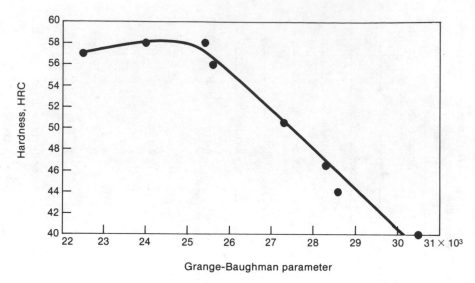

The steel was hardened by austenitizing at 1050 °C (1925 °F) followed by quenching.

**Fig. 9.7. Correlation of short-time time-temperature-hardness
data for tempering of type 440C martensitic stainless steel using
the Grange-Baughman parameter (Ref 115)**

temperature ranges listed in Table 9.2. This is done to minimize the time required,
as with induction tempering of lower-alloy steels. However, as the tempering
temperature is increased, extra care must be exercised in process control because
of the rapid drop in hardness with small temperature increases at high tempering
temperatures.

There is relatively little information on the time-temperature relationships that
should be used to obtain equivalent results in furnace and induction tempering
treatments of martensitic stainless steels. One such relationship, for type 440C
martensitic stainless steel austenitized and hardened to 60 HRC, is shown in
Fig. 9.7. These results are for treatments lasting between 10 and 420 s conducted
at temperatures of 540, 650, and 760 °C (1000, 1200, and 1400 °F) using a
high-speed resistance heating technique. Using the Grange-Baughman correlation
(Ref 28), it appears that the tempered hardnesses are single-value functions of the
tempering parameter.

HEAT TREATMENT OF TOOL STEELS

Like the low-alloy quenched-and-tempered steels, most tool steels can be hardened and tempered to produce materials with high strength and good wear resistance and toughness. Some of the more important classes of tool steels are the
following:

- Water-hardening tool steels ("W" classification)
- Shock-resisting tool steels ("S" classification)

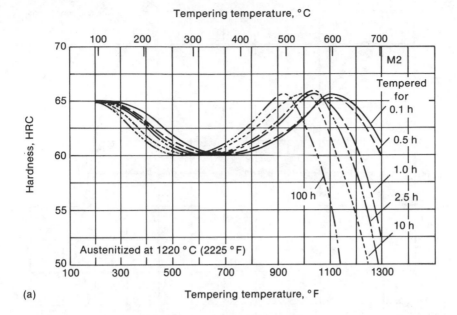

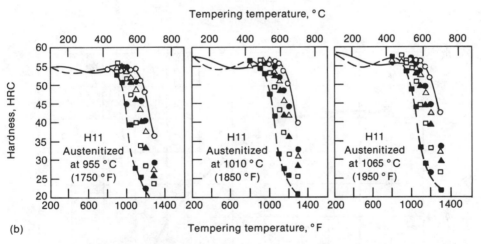

Symbols in (b): tempered for ○ 0.1 h; ● 0.5 h; △ 1.0 h; ▲ 2.5 h;
□ 10.0 h; ■ 100.0 h.

Fig. 9.8. Effects of temperature and time on tempered hardness of (a) M2 and (b) H11 tool steels (Ref 117)

- Oil-hardening cold work tool steels ("O" classification)
- Air-hardening cold work tool steels ("A" classification)
- High-carbon, high-chromium cold work tool steels ("D" classification)
- Hot work tool steels ("H" classification)
- Tungsten high speed tool steels ("T" classification)
- Molybdenum high speed tool steels ("M" classification).

As their name implies, tool steels are used in such applications as forging and extrusion dies and metalcutting tools. Because of their high alloy content and low

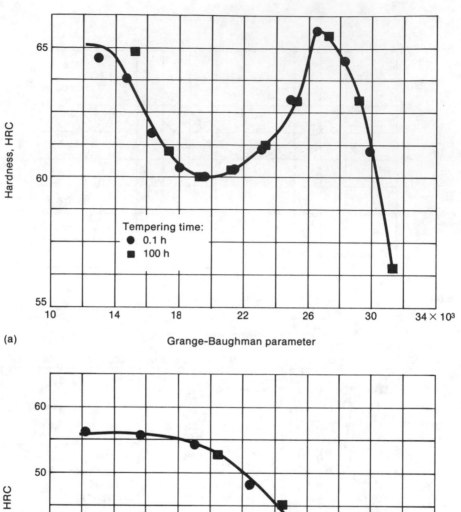

(a)

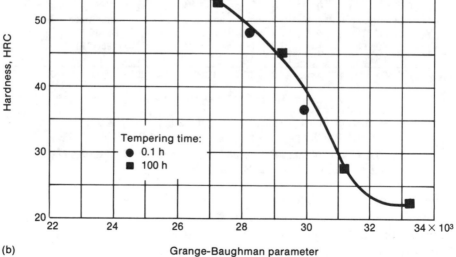

(b)

(a) M2 tool steel. (b) H11 tool steel austenitized at 955 °C
(1750 °F).

**Fig. 9.9. Tempering data from Fig. 9.8 replotted in terms of the
Grange-Baughman parameter**

thermal conductivity, tool steels are very susceptible to thermal shock during heating for austenitization or tempering. Thus, if induction is to be used in the hardening process, low generator frequencies and rather low power densities must be employed, particularly for through-hardening.

An example of induction through-hardening and tempering of tool steels is that described by Marschall (Ref 116). In this work, 5.1-cm (2.0-in.) diameter bars of D2 and H13 tool steels were heat treated by a vertical induction scanning method. Very uniform as-hardened and hardened-and-tempered hardnesses were obtained using a 1920-Hz generator and a power density of approximately 0.075 kW/cm^2 (0.48 kW/in.^{2}) for austenitizing, and 0.030 kW/cm^2 (0.19 kW/in.^{2}) for tempering. The austenitizing temperature was about 1095 °C (2000 °F), which is 55 to 85 °C (100 to 150 °F) higher than that used for furnace hardening treatments. The higher austenitizing temperatures for induction austenitizing of tool steels are required to dissolve more rapidly the alloy carbides present in the starting microstructure. By this means, austenitizing time for D2 and H13 was cut by a factor of about 5 to 10 (6 min for induction versus 30 to 60 min for furnace austenitizing). Induction tempering was achieved using heating times of 4 min.

Data for the above tool materials and others suggest that short-time induction tempering treatments can be devised. As examples, tempering data for M2 and H11 tool steels are shown in Fig. 9.8. For M2 high speed tool steel (Fig. 9.8a), the tempering curves are shifted to the right (i.e., to high temperatures) as the tempering time is lowered, a trend remarked upon previously for low-alloy steels. This behavior is followed for the secondary hardening peak as well as the softening response at higher tempering temperatures. Correlation of these tempering data using the Grange-Baughman parameter meets with good success. For example, a replot of the 0.1- and 100-h results in Fig. 9.8(a) in terms of the Grange-Baughman parameter leads to the graph in Fig. 9.9(a). The tempering correlation is seen to be quite good. The tempering data for H11 steel (Fig. 9.8b) can also be reduced to unique trend lines using the Grange-Baughman parameter. An example, for material austenitized at 955 °C (1750 °F), is given in Fig. 9.9(b).

Chapter 10

Cost Analysis and Future Developments in Induction Heat Treatment

As has been discussed at length in previous chapters, induction heat treatment of steels falls into two major categories: surface-hardening applications and through-hardening applications. In the former, the selection of induction methods rarely involves cost considerations. This is because of the need for special or accurately controlled hardness distributions and properties which can be achieved only by induction. For instance, hard and wear-resistant surfaces (under a state of compressive residual stress) and ductile cores are needed in parts which require a blend of strength, toughness, and fatigue resistance. High production rates, low labor and maintenance costs, the ability to use inexpensive carbon steels, and the need for little or no machining all serve to make induction surface hardening an attractive process from an economics viewpoint.

In contrast to surface hardening, through-hardening and tempering by induction methods may have to compete with other heat treating methods on the basis of cost. Some of the more important factors in justifying one or another method are the following:

1. Cost of equipment
2. Cost of fuel/electricity
3. Efficiency of equipment
4. Scale losses in process
5. Scrap losses in process
6. Labor requirements to operate process and for inspection
7. Labor requirements for maintenance
8. Floor space required for equipment
9. Subsequent operations required for process (e.g., straightening, machining, grinding, etc.)
10. Production rates and ability to eliminate handling operations and to use automated equipment.

Estimating the costs of items 1 through 8 is a relatively straightforward matter. On the other hand, the costs of items 9 and 10 are greatly dependent on the type of part being processed and on the other steps that are involved in its manufacture. These other operations may dictate whether a continuous operation with in-line heat treatment is feasible and whether an automated system with microprocessor controls can be employed. Nevertheless, induction can often lead to cost reductions under items 9 and 10 because of the ability to control distortion, to avoid scaling, and to use automated equipment. Consideration of the other cost factors will be discussed in the following sections. Here, the cost of induction hardening and tempering will be compared with the costs of more conventional means of heat treating steel — namely, those involving gas-fired furnaces.

COST FACTORS FOR INDUCTION VERSUS GAS-FIRED FURNACE HEAT TREATMENT OF STEEL

Cost of Equipment

The cost of induction generators is usually several times that of gas-fired furnaces for comparable heating capacity. For through-heating applications, single-frequency generators typically cost from two to two-and-one-half times as much as comparable furnaces. Dual-frequency generators, which are increasing in popularity nowadays, usually cost three times as much as equivalent furnaces. These differences in cost are based on furnaces without recuperators — devices which recover some of the energy lost with the flue gases. With recuperators, combustion air for the furnace is preheated by having outgoing flue gases transfer a part of their heat while flowing through a chamber through which the inlet air supply passes. Recuperators are classified according to whether the flue gas flows parallel to, against, or across the flow of combustion air, and are thus termed "parallel-flow," "counter-flow," and "cross-flow" recuperators; counter-flow types allow the highest preheat temperatures to be achieved. Furnaces equipped with recuperators have higher over-all efficiencies, and, because of the added construction expense, they cost about one-half as much as comparable induction equipment.

The absolute cost of induction generators varies with frequency and power rating. For a given power rating, it increases with frequency, and for a given frequency, the cost scales linearly with the power rating. In terms of 1983 dollars, typical costs range from $100 to $200 per kilowatt for 60-to-1000-Hz generators, from $200 to $300 per kilowatt for 3-to-10-kHz generators, and from $300 to $400 per kilowatt for radio-frequency generators. Since most through-heating applications require 10 kHz or less, $150 to $250 per kilowatt is a good figure to use in estimating capital equipment costs.

A typical gas-fired furnace in 1983 cost approximately $6000 to $7000 for each 1,000,000 Btu/h of capacity. This corresponds to about $20 to $25 per kilowatt. Frequently, furnaces are only about one-fourth as efficient as induction heating systems; a furnace comparable to an induction unit would cost about $80 to $100 per kilowatt.

Furnaces and induction equipment both have very long service lives. Many furnaces, built decades ago, although refurbished and relined with new refractory material many times, are still in use. Induction equipment usually has an equally long life. Because of this, amortization of either type of equipment is readily spread over a period of five to ten years or even longer.

A related cost in the equipment area concerns the amount of floor space needed for furnaces and induction generators. The former are generally much larger than the latter, and thus the larger physical plant needed to house furnaces should be considered in estimating capital equipment costs.

Fuel Costs

Unlike capital equipment expenses, the costs of fuel needed for induction and gas-fired furnaces are increasing as a result of increasing demand and diminishing supply. Electricity for induction heat treatment tends to be much more expensive than gas, although gas prices are increasing at a greater rate. Part of this higher price is a result of the fact that substantial energy losses are involved in the generation and transmission of electricity. Thus, the power supplied "at the plug" is only a fraction of that which could be obtained from the burning of fossil fuels or operation of a nuclear reactor under 100% efficiency at the power plant. However, price projections for electricity suggest a small annual increase (of only 1%). This is primarily due to favorable generation capacity levels and the reliance on coal as a fuel source by many electric utilities.

In the United States, the costs of electricity and gas vary a great deal from one region to another. For example, in 1983, the cost of electricity ranged from approximately 1.5¢ per kilowatt-hour in the Northwestern states, where hydro-electric power is abundant, to a high of about 12¢ per kilowatt-hour in the Northeast. In Midwestern states such as Michigan, Ohio, Pennsylvania, and Illinois, where perhaps the largest amount of metal processing takes place, this cost averaged approximately 6¢ per kilowatt-hour in 1983. The cost of natural gas does not show quite as much regional variation as that of electricity. Gas is most expensive in the Mid-Atlantic, Pacific, and Northeastern states (approximately $4.00 per million Btu*) and cheapest in the South (approximately $2.75 per million Btu). In the Midwestern regions, where gas is used most for metal processing, the cost to industrial users is about $3.50 per million Btu. In terms of electrical energy units, this last figure is equivalent to about 1.2¢ per kilowatt-hour, or one-fifth the cost of electricity.

Efficiency

The above cost differences between electricity and gas are mitigated to a large extent by the usually greater efficiencies of induction-based heat treating processes. The over-all efficiency of an induction system is dependent on the efficiency of the generator (termed the "terminal efficiency") and that of the coil. The

---

*The cost of gas is usually quoted in terms of heat content, because the content per unit volume of gas varies with the exact type of gas used.

Table 10.1. Comparative efficiencies of various power sources (Ref 40)

Power source	Frequency	Terminal efficiency, %	Coil efficiency, %	System efficiency, %
Supply system	50 to 60 Hz	93 to 97	50 to 90	45 to 85
Frequency multiplier	50 to 180 Hz	85 to 90	50 to 90	40 to 80
	150 to 540 Hz	93 to 95	60 to 92	55 to 85
Motor-generator	1 kHz	85 to 90	67 to 93	55 to 80
	3 kHz	83 to 88	70 to 95	55 to 80
	10 kHz	75 to 83	75 to 96	55 to 80
Static inverter	500 Hz	92 to 96	60 to 92	55 to 85
	1 kHz	91 to 95	70 to 93	60 to 85
	3 kHz	90 to 93	70 to 95	60 to 85
	10 kHz	87 to 90	76 to 96	60 to 85
Radio-frequency generator	200 to 500 kHz	55 to 65	92 to 96	50 to 60

terminal efficiency is equal to the ratio of the power output of the generator to the power input, and is typically fairly high, except in the case of radio-frequency generators. As discussed by Davies and Simpson (Ref 40), the terminal efficiency is often on the order of 85 to 95% (Table 10.1). It varies depending on the power rating, cooling system, and equipment. As described in previous chapters, coil efficiency depends on the design of the coil and its coupling to the workpiece as well as on workpiece properties. Although there are standard coil-design procedures for given part geometries, it is not unusual to redesign them once or several times for a specific application. For heat treating of steel below the Curie temperature (as in tempering), good coil designs allow energy-transfer efficiencies of 90 to 95%. In hardening treatments (above the Curie point), coil efficiencies are usually somewhat lower, averaging around 65 to 85%. However, this can be partially compensated for by dual-frequency generators in which the higher frequency is used above the Curie temperature. The net efficiency of the over-all induction system is therefore in the range of 55 to 85% (Table 10.1).

The efficiencies of gas-fired furnaces are much lower than those of induction systems, ranging between 10 and 25% depending on furnace design and operation. The heat generated by burning of the fuel in a furnace is transferred by the following means:

- Thermal energy transferred to the charge in the furnace
- Heat radiated through doors, slots, and other openings in the furnace
- Heat lost by transport of the products of combustion or by incomplete burning of the fuel
- Heat radiated through the walls and bottom of the furnace.

The thermal efficiency of a furnace is equal to the thermal energy transferred to the charge divided by the heat content of the fuel. Typical furnace efficiencies as a function of furnace temperature are shown in Fig. 10.1. These efficiencies are much lower than maximum theoretical values because of the heat losses cited above. Moreover, at low furnace temperatures, hotter combustion products must

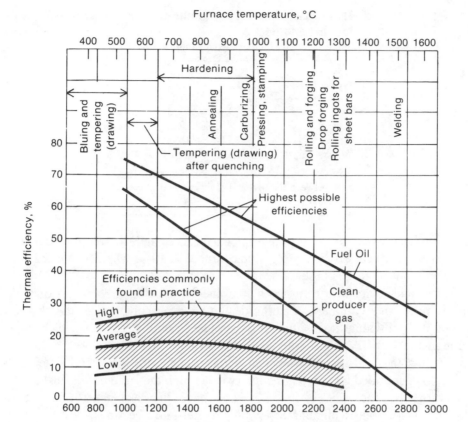

Furnace temperature, °C

Typical heat treating and metal fabrication temperature regimes are
indicated at top of graph.

**Fig. 10.1. Ideal and actual thermal efficiencies of batch-type in-
dustrial furnaces as a function of furnace temperature (Ref 118)**

be *cooled* prior to coming into contact with the charge in order to avoid overheating it. Because of this added drawback, efficiencies at low temperatures are
substantially lower than theoretical estimates.

The radiation and convection losses cited above vary with furnace type. Some
of the more common types of furnaces for heat treatment are known by the names
"box," "slot," "pusher," and "rotary hearth." A box furnace consists of an external
steel shell lined with an insulating refractory material, a heating system, and one
or more access doors to the heating chamber. Material is loaded into the chamber,
the doors are closed, and the material is then heated and cooled in a batch-type
process. A slot furnace is similar except that a long, narrow slot or furnace opening
is made by raising the furnace door. This slot is only of sufficient height to permit
the maximum size of stock being handled by the furnace to pass through the
opening. In this way, billets or finished products can be continually charged or
removed from the furnace. Because of this opening, however, convection and

Table 10.2.   Thermal efficiencies of fuel-fired furnaces
Source: Forging Industry Association, Cleveland.

Type of furnace	Thermal efficiency, %		
	Low	Average	High
Box	3.1	10.3	28.0
Slot	2.9	7.9	14.3
Pusher	13.0	18.5	21.5
Rotary hearth	13.2	(a)	46.5

(a) No explanation is given for the wide difference between the low and high values of thermal efficiency.

radiation heat losses are greater than in box furnaces and, therefore, efficiencies are lower. A typical efficiency for a box furnace is 10%, whereas that for a slot furnace is only 5 to 8% (Table 10.2).

Pusher and rotary-hearth furnaces combine some of the advantages of sealed box furnaces in a continuous heating design. In a pusher furnace, the charge is loaded into individual trays which are then placed on a larger tray which moves through the furnace. This larger tray fits through an opening in the furnace, eliminating the heat losses associated with an open slot. The total time in the furnace is controlled by varying the interval at which the large tray is pushed. For such furnaces, efficiencies of 15 to 20% are typical. In a rotary-hearth furnace, the charge is loaded onto a circular table which rotates into a heating chamber. By this means, the area of the inlet doors, and thus heat losses, can be minimized. Because of this design, the efficiencies of rotary-hearth furnaces can readily exceed those of the other furnaces discussed (Table 10.2).

To summarize, the average efficiency of an induction system is 60 to 85%, whereas the efficiency of an average furnace is only about 10 to 25%. Therefore, the four- or fivefold difference between the costs of electricity and gas can be compensated for by a nearly equivalent increase in efficiency afforded by induction.

Not included in the above discussion is the fact that the efficiency of a furnace or induction system may vary with temperature. For either type of equipment, efficiency generally decreases with temperature. This loss can be minimized in induction setups by use of dual-frequency generators. Such an alternative is not possible with gas furnaces because of the increases in radiation losses and flue-gas losses that occur as the temperature is increased. Because of these losses, it is sometimes preferable to carry out heating and heat treating operations using a furnace for preheating and induction equipment for final heating.

One final consideration in determining the over-all efficiency of a furnace system is the number of shifts that are to be worked. If the furnace is to be operated sporadically—i.e., for only one or two shifts—substantial energy is wasted during cooling and reheating of the furnace. The loss of energy can be controlled to a certain extent by letting the furnace "idle" at some intermediate temperature when not in use. The decrease in over-all efficiency in such instances has no effect, of course, in shops in which induction is employed. Here, full heating production can be started and stopped at will with no loss in over-all thermal efficiency.

Scale and Scrap Losses

One of the more important factors in selection of steel heating equipment is the amount of material lost due to scaling and scrap. Such a consideration is especially important in high-production shops in which the cost of the steel can constitute as much as one-half of the net cost of the finished product.

There are no firm figures on scale and scrap losses in tempering operations. However, it is unlikely that such losses are sizeable in view of the low temperatures used for this process. By contrast, scale losses in the hardening operation can be very substantial. In furnace heating of steel billets to forging temperatures (about 1200 °C, or 2200 °F), scaling amounting to 2 to 4% of the gross weight is not unusual. At austenitizing temperatures (about 900 °C, or 1650 °F), the amount of scale formation is probably somewhat less (perhaps 1 to 2%), depending on the exact temperature and heating time used. Indirectly, the scaling problem entails the additional costs of machining the surfaces of the hardened part.

Because of its rapidity, induction hardening typically gives rise to little if any scaling. A figure of 0.5% is often used for scaling losses for steel heated to forging temperatures. For material heated only to austenitizing temperatures, this number is likely to be somewhat less.

In a related area, scrap losses due to improper heat treatment are usually higher in gas-fired furnace operations than in induction ones. In furnace-based processes, thermal cracking (during heating) and distortion (during quenching) are the usual causes of part rejection. Such defects cause scrap losses on the order of 1 to 2%. In induction operations, however, the above problems can be minimized by part rotation and proper design of the quench system, among other things. Scrap losses associated with induction heating of steel billet stock for forging are usually around 0.25 to 0.5%. It is not unreasonable to assume similar losses in induction hardening and tempering.

Labor Costs

Induction systems require less labor for operation and maintenance than comparable furnace-based heat treating systems. Reliable solid-state power generators are now coming into wide use, and the fast response and ease of controllability of induction heaters make them very amenable to automation. Automation also allows the use of less-skilled operators with a minimal amount of training. The use of automatic as opposed to manual operation in a single-shot induction application has reduced labor requirements by as much as 50 to 80% in some instances (Ref 63). On the other hand, furnace heat treating processes normally require a considerable amount of labor for loading, unloading, and handling of the steel to be processed. In addition, maintenance costs for furnaces tend to be higher because they need to be relined intermittently. To take these factors into account, a general rule of thumb for cost-analysis purposes is that induction heat treating usually requires one-half the labor needed for furnace processes with an equivalent production rate.

COST ANALYSES FOR INDUCTION VERSUS GAS-FIRED FURNACE HEAT TREATMENT OF STEEL

The data cited previously in this chapter can be used to obtain an estimate of the cost of induction versus gas-furnace processes. Cost analyses of this sort for heating of billets for forging have been quoted several times in the literature (Ref 40 and 119). An example of such an analysis is given in Table 10.3, which compares an induction system operating at 52% efficiency and a gas-fired furnace operating at 17.5% efficiency. Neglecting the capital investment, it can be seen that the only item for which induction poses a greater expense is energy. With the more rapid increase in gas prices since 1980, the most recent year for which costs are estimated in the table, the differential has decreased somewhat. It can also be noted that substantial savings in labor and in scrap and scale losses are realized through the use of induction, leading to smaller over-all cost. Such lower costs can aid in the amortization of the induction equipment over an equal or shorter period of time in a high-production shop.

As another example, consider the hardening of 1045 steel parts using induction and gas-fired furnaces. We shall assume that the production rate is 20 tons per

Table 10.3. Comparison of costs for induction and natural-gas heating of billets for forging (Ref 119)

Item	Cost per ton of product 1977	1980
Induction heating		
Electric supply energy	$ 8.34	$12.70
Production labor	0.93	1.25
Maintenance labor	0.33	0.44
Scale loss	1.75	2.33
Scrap loss	1.75	2.33
Total	$13.10	$19.05
Natural-gas heating		
Energy	$ 5.93	$ 9.20
Production labor	1.87	2.49
Maintenance labor	0.67	0.89
Scale loss	7.00	9.34
Scrap loss	7.00	9.34
Total	$22.47	$31.26

hour, that the plant operates two shifts or 4000 hours per year, and that the cost of raw material is $700 per ton; however, scrap (and scale) will be assumed to cost somewhat more than this ($800 per ton) because of the value added during processing. Scrap and scale rates will each be assumed to be 1.5% for furnace operations and 0.5% for induction treatments. Other costs will be taken to be $15 per hour (including benefits) for labor, 6¢ per kilowatt-hour for electricity, and $4 per million Btu for gas.

The cost breakdown for this heat treatment operation, neglecting raw-material costs, which would be the same for both processes, and assuming equipment amortization over a five-year period, is listed in Table 10.4. Austenitizing temperature is taken to be 900 °C (1650 °F), and the induction system to be purchased is assumed to operate at an over-all efficiency of 60%. For the throughput desired, the induction generator to supply the 5300 kW needed to heat the steel would cost approximately $1.3 million. A comparable furnace with an assumed efficiency of 15% would cost roughly one-third this amount, or $450,000. The annual energy cost for the induction process is based on 4000 hours of operation, and that for the furnace is based on 6000 hours since this equipment must be idled even when not on a production run. Because of the low efficiency of the furnace and the need for it to be run at all times, the total yearly energy cost is *greater* than that for the induction system. Scale, scrap, and labor costs also tend to favor the induction process. Summing all of these costs, it can be seen that induction is much cheaper than furnace hardening.

If similar equipment is used for a through-tempering process, relative costs can be estimated from the data in Table 10.4. This is accomplished by assuming negligible scale and scrap losses. Thus, neglecting changes in energy costs (which would be lower because of lower temperatures for tempering), the yearly costs for induction and furnace tempering would be approximately $1.635 million and $2.02 million, respectively, and the two processes would be competitive. With the

Table 10.4.   Comparison of costs for induction and furnace hardening of 1045 steel products

| | Cost (in thousands of dollars) | | | |
| | Induction hardening | | Furnace hardening | |
Item	Initial	Annual(a)	Initial	Annual(a)
Capital equipment........	1300	265	450	90
Energy.................	...	1280	...	1750
Scale loss	...	320	...	960
Scrap loss	...	320	...	960
Labor(b)	...	90	...	180
Total annual cost.................		2275		3940

(a) At a production rate of 20 tons per hour for 2 shifts per day, and assuming amortization over a 5-year period and neglecting interest.
(b) Assuming 1½ men for induction unit and 3 men for furnace operation (includes operation and maintenance).

increasing cost of labor and the more rapid increase in the cost of gas relative to electricity, however, economics will probably strongly favor induction in the future.

To estimate the impact of equipment efficiency and fuel costs on total heat treating costs, a cost-sensitivity analysis can be performed. For the austenitizing example discussed above and summarized in Table 10.4, such an analysis leads to the dependence of *total* fuel costs on equipment efficiency and unit fuel cost shown in Fig. 10.2. From this plot, it can be seen that the energy cost of a gas-furnace operation can be cut by about $1 million (from $1.75 million to $0.75 million) by increasing the furnace efficiency to about 25%. This would lower the yearly operating cost of the furnace process to approximately $3 million. Similarly, a decrease in induction heating efficiency from 60 to 50% and relatively moderate increases in the cost of electricity could readily increase over-all energy costs from $1.28 million (Table 10.4) to approximately

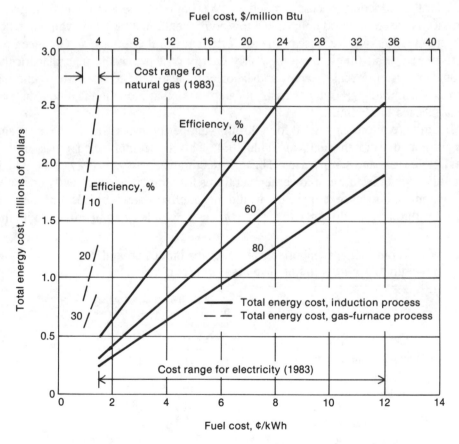

Dependence of energy cost on both system efficiency and fuel cost is shown for induction and furnace-based processes.

Fig. 10.2. Total energy cost for hardening operation described in Table 10.4

$2 million, resulting in a total annual cost of about $3 million for the induction process as well. This would make the induction and furnace hardening operations very competitive.

The above examples are all for shops with high production rates. In these instances, the cost of raw materials forms a sizeable proportion of the over-all cost. For instance, neglecting capital equipment costs, the cost per ton for the induction hardening process in Table 10.4 is only $25, whereas the raw-material cost was $700 per ton. Furthermore, in shops where production rates are not very high, as in job shops, the cost advantage enjoyed by induction hardening may not be present because of the high labor costs associated with design, construction, and implementation of new coils. In these instances, furnace heat treatment will probably remain more cost-effective.

COSTS OF ELECTRIC FURNACE HEAT TREATMENT

Although attention has been focused in this chapter on the cost of induction versus gas-fired furnace heat treatment of steel, electric furnaces find wide application in this area of metal processing also. The most common type of electric furnace used for these purposes is the resistance-type furnace with silicon carbide or metallic (usually nickel-chromium or nickel-chromium-iron alloy) heating elements. An electric current is passed through the elements, giving rise to I^2R heating, which is transferred to the workpiece via radiation or forced-air convection. Among the advantages of electric furnaces are the following:

- Accurate and responsive automatic temperature control; good temperature uniformity within the furnace, often better than that in gas-fired furnaces
- Compactness and simplicity of construction
- Cleanness (no combustion products requiring disposal)
- Low heat loss to surroundings
- Good adaptability to controlled-atmosphere processing.

Some of the disadvantages include the generally greater difficulty in repair of electric furnaces and the need for more stringent maintenance as compared with gas-fired furnaces. The consequences of neglect in maintenance can be much more serious (e.g., burnout of resistance heating elements) in this type of furnace as well. Another disadvantage, when compared with induction heat treatment setups, is the need to "idle" electric furnaces when not in use at or near heat treating temperatures in order to avoid costly reheating starting at ambient or slightly higher temperatures.

Because they lose no heat as a result of the exit of hot flue gases as in gas-fired furnaces, electric furnaces tend to have high efficiencies; usually around 60% of the resistive heat given off by the heating elements may be absorbed by the charge. When compared with the efficiencies obtained in typical induction heat treatments, those for electric furnaces (Fig. 10.3) make them quite competitive from an energy standpoint. Thus, many tempering operations utilizing electric power

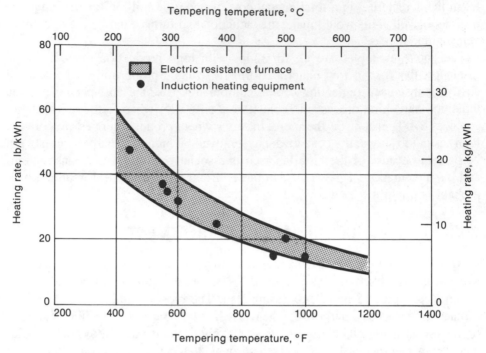

Note the equivalence of the efficiencies of the two processes.

Fig. 10.3. Required heat input as a function of tempering temperature for tempering of steel products using induction and electric-furnace equipment (Ref 50)

are performed in furnaces rather than by means of induction. This is particularly true for treatment of very heavy cross sections (i.e., over 12.7 cm, or 5 in.) requiring very uniform properties, for which through-tempering by induction may not be suitable.

FUTURE OUTLOOK

Because of its many advantages, induction heating is capturing a large share of the steel heat treating market despite the high costs of equipment and electricity. Conventional noninduction hardening methods will continue to be used for complex geometries and small production runs. However, there appears to be some potential for induction to make a significant impact in nonhardening heat treating applications such as annealing and tempering.

The rapid increase in the cost of gas is undoubtedly playing a large role in the over-all economic picture. Research and development, primarily by equipment manufacturers, is also opening new markets for induction heat treatment. This research falls into several major categories, among which are the following:

● *Improvements in Equipment.* Work is under way to increase efficiency and productivity of induction installations. The use of dual-frequency generators and solid-state semiconductor units with continuously variable frequency are among the developmental thrusts being pursued in this area. The incentive to develop higher-power, higher-frequency ($\geq$50 kHz) solid-state power supplies is great. These generators would be more efficient (approximately 90%) and cheaper than the vacuum-tube units (about 450 kHz) currently in use and could be used to perform many of the surface heat treating operations for which the latter are often employed. Concurrently, attempts at developing improved sensors for use with microprocessors to control variable-frequency units are under way. Also, means of incorporating several heating or heat treating, forming, and machining steps using programmable controllers and flexible manufacturing systems appear to be on the horizon insofar as over-all process developments are concerned.

- *Improvements in Coil Design.* Through-heating (and heat treating) of complex geometries and thin materials is a problem for conventional induction techniques due to nonuniformity of the induced eddy-current distributions. Improved coil designs and the introduction of transverse flux induction heating have shown that substantial increases in efficiency are possible in heat treating parts previously restricted to furnace techniques.

- *Modeling.* Analytical and physical modeling of induction heating processes represents an important research area which may help in the development of improved coil designs and process controls. Modern numerical analysis techniques allow computer simulation of the complicated thermal and electromagnetic interactions between the coil and the workpiece, permitting a more rational design for a broad range of applications. Physical modeling, in which the coil and workpiece are scaled down and the frequency is increased, is being used more and more. By this means, temperature distributions in a small workpiece similar to those in a full-scale application are developed, enabling the testing of nonstandard geometries at considerably reduced cost compared with full-scale testing.

- *Improvements in System Efficiency.* Improvements in the over-all efficiency of induction heating systems are being approached from the standpoint of heat recovery and the use of hybrid (partially fuel-fired) systems. Because the heat contained in the coil cooling water is at a temperature too low to be used effectively in most production operations, heat pumps will have to be employed to increase the temperature to a practical level. The thermodynamic and economic balance between cooling-water temperature, coil efficiency, and heat-pump energy requirements is being carefully examined to make such systems viable. Similarly, hybrid systems, in which billets are preheated in a combustion furnace and then brought to temperature in an induction heater, are undergoing economic evaluation in order to optimize their use in various applications.

Appendix A

Conversion Factors

Various conversion factors are often used in heat treating practice, be it by induction or furnace-based techniques. Some of the most important ones are as follows:

I. *Length and area*
   1 metre = 3.281 feet = 39.37 inches
   1 square metre = 10^4 square centimetres = 10.76 square feet
   = 1550 square inches

II. *Force, mass, and density*
   1 newton = 10^5 dynes = 0.1020 kilogram force = 0.2248 pound
   1 kilogram = 1000 grams = 2.205 pounds mass
   1 kilogram/metre^3 = 0.001 gram/centimetre^3 = 3.615 × 10^{-5} pound/inch^3

III. *Energy*
   1 joule = 1 newton-metre = 1 watt-second
   1 joule = 0.7376 foot-pound = 2.778 × 10^{-7} kilowatt-hour
   = 0.2389 calorie = 9.481 × 10^{-4} Btu

IV. *Power*
   1 joule/second = 1 newton-metre/second = 1 watt
   1 watt = 0.7376 foot-pound/second = 0.2389 calorie/second
   = 3.413 Btu/hour

V. *Temperature* (°C ≡ Celsius, °F ≡ Fahrenheit, K ≡ Kelvin, °R ≡ Rankine)
   °C = 5(°F − 32)/9
   K = 5(°R)/9
   °C = K − 273
   °F = °R − 460

VI. *Thermal properties*
   Specific heat: 1 calorie/gram · °C = 1 Btu/pound · °F

Thermal conductivity: 1 calorie/second · centimetre · °C
    = 241.8 Btu/hour · foot · °F = 418.4 watts/metre · K
Thermal diffusivity: 1 centimetre^2/second = 0.155 inch^2/second
Heat transfer coefficient: 1 calorie/second · centimetre^2 · K
    = 7.369 × 10^3 Btu/hour · foot^2 · °F = 4.184 watts/centimetre^2 · K

VII. *Pressure, stress*
    1 pascal = 1 newton/metre^2
    10^3 pascal = 1 kPa = 0.145 pound/inch^2
    10^6 pascal = 1 MPa = 0.145 ksi (0.145 × 10^3 pounds/inch^2)

VIII. *Hardness* (see Table A.1.)

Table A.1.　Approximate equivalent hardness numbers for steel (Ref 120)

These values are for carbon and alloy steels in the annealed, normalized, and quenched-and-tempered conditions; they are less accurate for steels in the cold worked condition and for austenitic steels. The values in **boldface type** correspond to the values in the joint SAE-ASM-ASTM hardness conversions as printed in ASTM E140, Table 2. The values in parentheses are beyond the normal range and are given for information only.

Rockwell C-scale hardness No.	Vickers hardness No.	Brinell hardness No., 3000-kg load, 10-mm ball Standard ball	Tungsten carbide ball	Rockwell hardness No. A scale, 60-kg load, Brale indenter	B scale, 100-kg load, 1/16-in.-diam ball	D scale, 10-kg load, Brale indenter	Rockwell superficial hardness No., superficial Brale indenter 15N scale, 15-kg load	30N scale, 30-kg load	45N scale, 45-kg load	Knoop hardness No., 500-g load and greater	Shore Scleroscope hardness No.	Tensile strength (approx), ksi	Rockwell C-scale hardness No.
68	940	...	...	85.6	...	76.9	93.2	84.4	75.4	920	97	...	68
67	900	...	...	85.0	...	76.1	92.9	83.6	74.2	895	95	...	67
66	865	...	...	84.5	...	75.4	92.5	82.8	73.3	870	92	...	66
65	832	...	(739)	83.9	...	74.5	92.2	81.9	72.0	846	91	...	65
64	800	...	(722)	83.4	...	73.8	91.8	81.1	71.0	822	88	...	64
63	772	...	(705)	82.8	...	73.0	91.4	80.1	69.9	799	87	...	63
62	746	...	(688)	82.3	...	72.2	91.1	79.3	68.8	776	85	...	62
61	720	...	(670)	81.8	...	71.5	90.7	78.4	67.7	754	83	...	61
60	697	...	(654)	81.2	...	70.7	90.2	77.5	66.6	732	81	...	60
59	674	...	(634)	80.7	...	69.9	89.8	76.6	65.5	710	80	351	59
58	653	...	615	80.1	...	69.2	89.3	75.7	64.3	690	78	338	58
57	633	...	595	79.6	...	68.5	88.9	74.8	63.2	670	76	325	57
56	613	...	577	79.0	...	67.7	88.3	73.9	62.0	650	75	313	56
55	595	...	560	78.5	...	66.9	87.9	73.0	60.9	630	74	301	55
54	577	...	543	78.0	...	66.1	87.4	72.0	59.8	612	72	292	54
53	560	...	525	77.4	...	65.4	86.9	71.2	58.6	594	71	283	53
52	544	(500)	512	76.8	...	64.6	86.4	70.2	57.4	576	69	273	52
51	528	(487)	496	76.3	...	63.8	85.9	69.4	56.1	558	68	264	51
50	513	(475)	481	75.9	...	63.1	85.5	68.5	55.0	542	67	255	50
49	498	(464)	469	75.2	...	62.1	85.0	67.6	53.8	526	66	246	49
48	484	(451)	455	74.7	...	61.4	84.5	66.7	52.5	510	64	238	48
47	471	442	443	74.1	...	60.8	83.9	65.8	51.4	495	63	229	47
46	458	432	432	73.6	...	60.0	83.5	64.8	50.3	480	62	221	46
45	446	421	421	73.1	...	59.2	83.0	64.0	49.0	466	60	215	45
44	434	409	409	72.5	...	58.5	82.5	63.1	47.8	452	58	208	44
43	423	400	400	72.0	...	57.7	82.0	62.2	46.7	438	57	201	43
42	412	390	390	71.5	...	56.9	81.5	61.3	45.5	426	56	194	42
41	402	381	381	70.9	...	56.2	80.9	60.4	44.3	414	55	188	41
40	392	371	371	70.4	...	55.4	80.4	59.5	43.1	402	54	182	40
39	382	362	362	69.9	...	54.6	79.9	58.6	41.9	391	52	177	39
38	372	353	353	69.4	...	53.8	79.4	57.7	40.8	380	51	171	38
37	363	344	344	68.9	...	53.1	78.8	56.8	39.6	370	50	166	37
36	354	336	336	68.4	(109.0)	52.3	78.3	55.9	38.4	360	49	161	36
35	345	327	327	67.9	(108.5)	51.5	77.7	55.0	37.2	351	48	157	35
34	336	319	319	67.4	(108.0)	50.8	77.2	54.2	36.1	342	47	153	34
33	327	311	311	66.8	(107.5)	50.0	76.6	53.3	34.9	334	46	149	33
32	318	301	301	66.3	(107.0)	49.2	76.1	52.1	33.7	326	44	145	32
31	310	294	294	65.8	(106.0)	48.4	75.6	51.3	32.5	318	43	141	31
30	302	286	286	65.3	(105.5)	47.7	75.0	50.4	31.3	311	42	138	30
29	294	279	279	64.7	(104.5)	47.0	74.5	49.5	30.1	304	41	135	29
28	286	271	271	64.3	(104.0)	46.1	73.9	48.6	28.9	297	40	131	28
27	279	264	264	63.8	(103.0)	45.2	73.3	47.7	27.8	290	39	128	27
26	272	258	258	63.3	(102.5)	44.6	72.8	46.8	26.7	284	38	125	26
25	266	253	253	62.8	(101.5)	43.8	72.2	45.9	25.5	278	38	122	25
24	260	247	247	62.4	(101.0)	43.1	71.6	45.0	24.3	272	37	119	24
23	254	243	243	62.0	100.0	42.1	71.0	44.0	23.1	266	36	117	23
22	248	237	237	61.5	99.0	41.6	70.5	43.2	22.0	261	35	114	22
21	243	231	231	61.0	98.5	40.9	69.9	42.3	20.7	256	35	112	21

Appendix B

Typical Microstructures Formed During Heat Treatment of Steel

The properties of all steels are intimately related to the microstructures developed by heat treatment. The various heat treatments which are commonly used for steels are briefly described in Chapter 2. In this appendix, a sampling of typical microstructures is presented. These illustrations are by no means all-inclusive, but are shown in order to give a general idea of the appearance of the constituents that may be observed during microscopic observation of polished and etched steel specimens. Thus, these pictures may be employed for at least a first-order quality evaluation of heat treating processes. The reader is referred elsewhere for more in-depth treatments of the development of microstructures in steels (Ref 121 and 122).

ANNEALED AND NORMALIZED MICROSTRUCTURES

The microstructures developed by annealing and normalizing are generally soft. Typical annealing treatments for steels include stress-relief and spheroidization types. In the former, little microstructural change occurs because of the low temperatures employed. On the other hand, spheroidization annealing is a heat treatment carried out at higher temperatures (just below the lower critical temperature, A_1) and leads to the softest and most ductile condition for a particular steel. During spheroidization annealing, as its name implies, carbides become globular or "spheroidized." This is illustrated in Fig. B.1 for a 4130 steel which has been hot rolled. Following hot rolling, the steel was air cooled to 675 °C (1250 °F) — a temperature about 50 °C (90 °F) below the A_1 temperature. At this temperature, the austenite which has not transformed during cooling decomposes to form the lamellar structure characteristic of pearlite (Fig. B.1a). When the steel is annealed or held at 675 °C (1250 °F) for an extended period of time, the carbide lamellae in the pearlite spheroidize, giving rise to the microstructure shown in Fig. B.1(b).

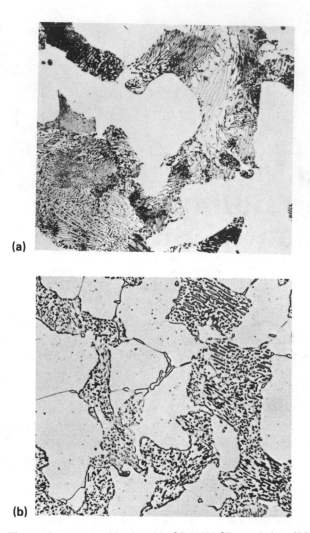

(a)

(b)

The steel was austenitized at 845 °C (1550 °F), cooled to 675 °C (1250 °F), held for either (a) 2 or (b) 112 h, and air cooled. The microstructures consist of (a) partially spheroidized pearlite and ferrite and (b) nearly totally spheroidized pearlite and ferrite. Nital etch. Magnification, 750×.

Fig. B.1. Microstructures of hot rolled 4130 steel (Ref 123)

The photographs in Fig. B.1 also show large white areas of ferrite. This ferrite was produced during air cooling at temperatures *between* the upper (A_3) and lower (A_1) critical temperature and is termed "proeutectoid" ferrite. Proeutectoid phases are also developed during normalizing. In normalization, the steel is austenitized and allowed to cool slowly, usually in air. Typical microstructures developed by such treatments are shown in Fig. B.2 and B.3. In the former figure, the white constituent in both instances is proeutectoid ferrite. The other, or lamellar, struc-

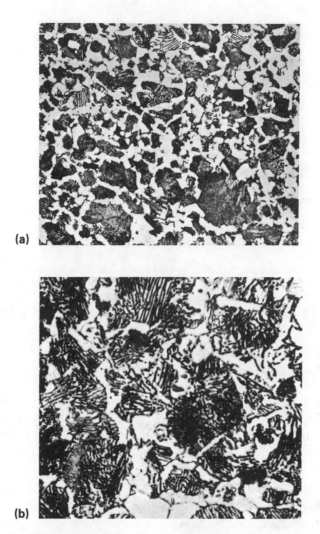

(a)

(b)

Lamellar pearlite (darker constituent) and ferrite (white areas) in
(a) 1045 and (b) 4140 steels produced by austenitizing and subsequent (a) air cooling or (b) furnace cooling to 620 °C (1150 °F)
followed by air cooling to room temperature. Picral etch. Magnifications: (a) 500×; (b) 1000×.

**Fig. B.2. Microstructures of normalized 1045 and 4140 steels
(Ref 124)**

ture is again pearlite formed by relatively slow cooling at temperatures below the
A_1 temperature. A similar microstructure, in which pearlite and a proeutectoid
phase of cementite have been formed, is depicted in Fig. B.3. Reference to the
iron-carbon phase diagram (Fig. 2.3) reveals that *hypoeutectoid* alloys
(C ≤ 0.8% in carbon steels) form proeutectoid ferrite, and that *hypereutectoid*
alloys (C > 0.8% in carbon steels) form proeutectoid cementite.

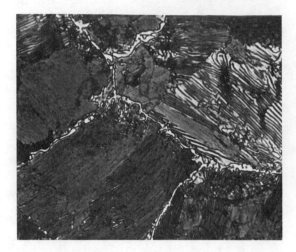

The steel was air cooled following hot rolling at approximately 1050 °C (1925 °F). The matrix is pearlite; cementite is found at the grain boundaries. Picral/nital etch. Magnification, 1000×.

Fig. B.3. Microstructure of normalized 52100 steel bar (Ref 125)

MICROSTRUCTURES FORMED BY ISOTHERMAL TRANSFORMATION OF AUSTENITE

As discussed in Chapter 2, a variety of microstructures may be produced by austenitizing followed by isothermal transformation below the A_1 temperature. At high subcritical transformation temperatures, pearlite is formed. At lower temperatures, bainite is formed. These steel morphologies are illustrated in Fig. B.4 through B.7.

The time dependence of pearlite formation for a eutectoid carbon steel is depicted in Fig. B.4. Following transformation for the indicated times, the steel samples were quenched, thereby causing martensite (which etches light in the photographs) to be formed from the remaining austenite. Thus, the amount of pearlite formed after the indicated times can be ascertained. Furthermore, such observations may be used to construct TTT diagrams such as those shown in Fig. 2.9(a), 2.11, and 2.12.

Isothermal transformation at lower subcritical temperatures leads to the formation of bainite. At temperatures around 450 °C (840 °F), a "feathery," or upper, bainite is observed. This is illustrated in Fig. B.5 for a eutectoid steel and in Fig. B.6 for various other carbon steels. In the former figure, the white-etching constituent is again martensite formed from untransformed austenite during quenching; Fig. B.5 also reveals the time dependence for bainite formation.

Isothermal transformation at temperatures below that at which feathery bainite is formed and above the martensite start, or M_s, temperature results in the development of lower bainite. This form of bainite is distinctly lenticular with a mottled rather than a feathery appearance. It is illustrated by the lower bainite formed in eutectoid steel at 300 °C (570 °F) in Fig. B.7.

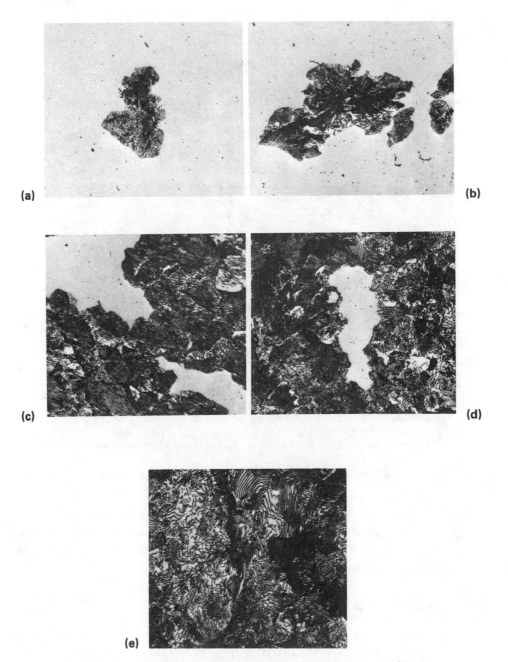

Pearlite formed in a eutectoid (1080) steel austenitized at 860 °C (1580 °F) and isothermally transformed at 705 °C (1300 °F) for (a) 150, (b) 300, (c) 600, (d) 800, or (e) 2000 s. The large white areas in (a) through (d) are martensite formed from untransformed austenite during quenching after isothermal transformation. Picral etch. Magnification, 250×.

Fig. B.4. Pearlite microstructures in 1080 steel (Ref 126)

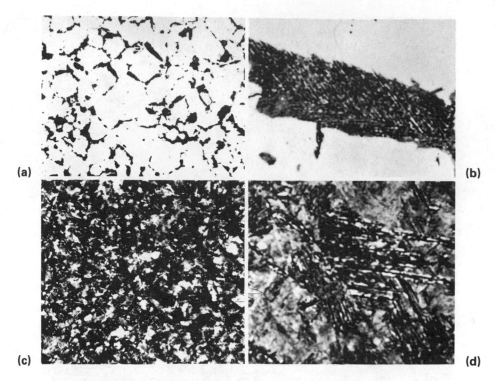

Upper ("feathery") bainite formed in a eutectoid (1080) steel austenitized at 860 °C (1580 °F) and isothermally transformed at 450 °C (840 °F) for (a, b) 0.5 and (c, d) 30 s. The white areas are martensite formed from untransformed austenite during quenching after isothermal transformation. Picral etch. Magnifications: (a, c) 250×; (b, d) 2000×.

Fig. B.5. Upper bainite microstructures in 1080 steel (Ref 127)

During *continuous* cooling, microstructures similar to those produced by isothermal transformation may be developed. Generally, these consist of combinations of pearlite and upper and lower bainite whose proportions and appearance depend on the specific cooling rate.

MARTENSITE AND TEMPERED MARTENSITE

If the steel is cooled sufficiently rapidly following austenitizing, the formation of pearlite and bainite is avoided, and martensite is produced. Martensite forms athermally (i.e., as the temperature is decreased) as opposed to isothermally (i.e., after a period of time at a fixed temperature). As discussed in Chapter 2, martensite formation begins at the M_s temperature and is completed at the M_f temperature.

Fully hardened, fully martensitic microstructures of various carbon steels are illustrated in Fig. B.8. For steels with carbon contents of 0.6% or lower, the martensite morphology is known as "lath." These laths consist of parallel arrays

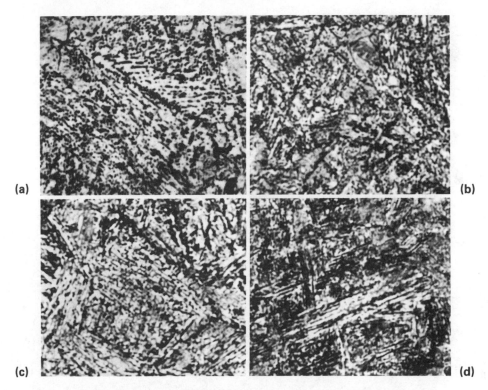

(a)

(b)

(c)

(d)

Upper ("feathery") bainite formed following austenitization and isothermal transformation at 450 °C (840 °F). Steels and isothermal transformation times were as follows: (a) 1015, 10 s; (b) 1025, 10 s; (c) 1040, 20 s; (d) 1060, 40 s. Picral etch. Magnification, 2000×.

Fig. B.6. Upper bainite microstructures in various carbon steels (Ref 128)

or packets of basic units which are 0.1 to 0.5 μm thick and which contain a high density of defects in the crystal structure known as dislocations. The individual laths are not resolvable in Fig. B.8; it is the packets of laths which are observed. At carbon contents of about 0.8% and higher, the martensite morphology is known as "plate"; it is lenticular in shape.

Tempering of martensite leads to noticeable microstructural changes. These are illustrated in Fig. B.9 for a 1040 steel. At low tempering temperatures and short tempering times, the carbide precipitation which occurs during tempering is not evident (e.g., Fig. B.9a to B.9h); it is too fine. However, the microstructure etches increasingly darker as the temperature or time is increased. At high temperatures and long times, particles of cementite, which are resolvable optically, are found (Fig. B.9i through B.9l and B.10).

The microstructure of tempered martensite, like the hardness, can be correlated using the tempering parameter discussed in Chapter 6. For example, tempering treatments of 700 °C (1290 °F) for 30 min and 600 °C (1110 °F) for 32 h result in

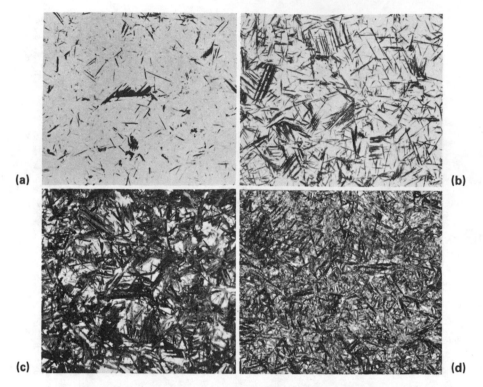

Lower bainite formed in a eutectoid (1080) steel austenitized at
860 °C (1580 °F) and isothermally transformed at 300 °C (570 °F)
for (a) 70, (b) 200, (c) 800, and (d) 2000 s. The white areas are
martensite formed from untransformed austenite during quenching
after isothermal transformation. Picral etch. Magnification, 250×.

Fig. B.7. Lower bainite microstructures in 1080 steel (Ref 129)

Grange-Baughman tempering parameters of 31,010 and 30,660, respectively. The
similarity of the tempered martensite microstructures in Fig. B.9(l) and B.10(d)
is, therefore, as expected.

MISCELLANEOUS MICROSTRUCTURES

Several other photographs are shown in Fig. B.11 and B.12 and are illustrative
of surface features in surface induction hardened steels and decarburized steels. In
Fig. B.11, the near-surface region of a surface-hardened 1050 steel is depicted.
The top portion of the photograph shows the quenched-and-tempered layer of
martensite; below it is the normalized starting microstructure of pearlite and
grain-boundary proeutectoid ferrite (similar in appearance to Fig. B.2a).

Figure B.12 shows a quenched-and-tempered 10B35 steel whose surface has
become decarburized. The substrate is martensitic, but the surface layer is one of
ferrite grains. The ferrite structure results from decarburization and, thus, the low
carbon content at the surface. Because of its low carbon content, the surface layer

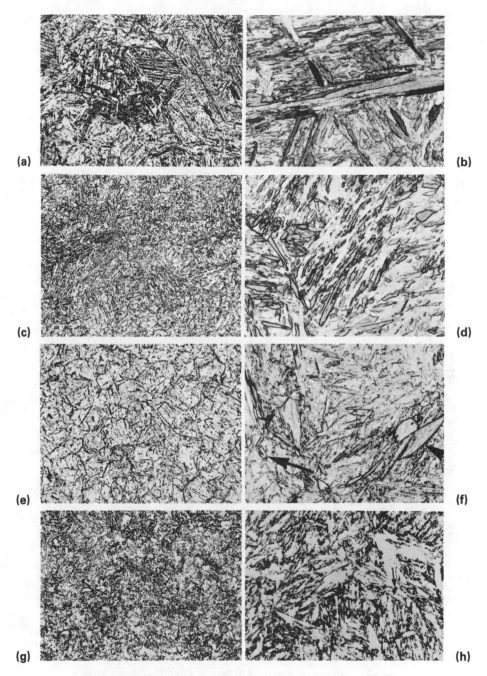

(a)   (b)

(c)   (d)

(e)   (f)

(g)   (h)

Martensite formed in austenitized and water quenched thin steel samples. Steels and austenitization temperatures were as follows: (a, b) 1025, 925 °C (1700 °F); (c, d) 1040, 875 °C (1605 °F); (e, f) 1060, 860 °C (1580 °F); (g, h) 1080, 860 °C (1580 °F). Nital etch. Magnifications: (a, c, e, g) 250×; (b, d, f, h) 1000×.

Fig. B.8. Fully hardened, fully martensitic microstructures in various carbon steels (Ref 130)

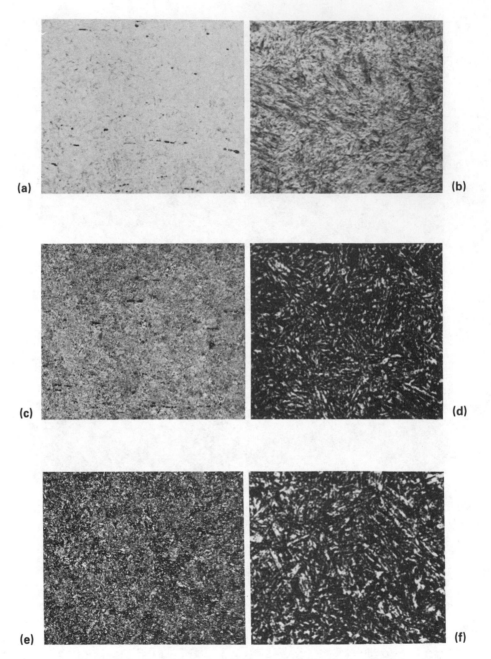

Samples were austenitized at 850 °C (1560 °F), water quenched, and tempered for 30 min at (a, b) 200 °C (390 °F); (c, d) 300 °C (570 °F); (e, f) 400 °C (750 °F); (g, h) 500 °C (930 °F); (i, j) 600 °C (1110 °F); (k, l) 700 °C (1290 °F). Picral etch. Magnifications: (a, c, e, g, i, k) 250×, (b, d, f, h, j, l) 1000×.

Fig. B.9. Tempered martensite microstructures in 1040 steel (Ref 131)

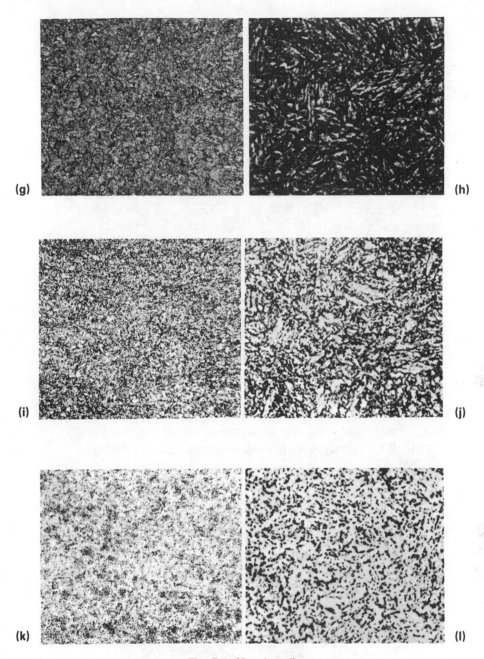

Fig. B.9. (Continued)

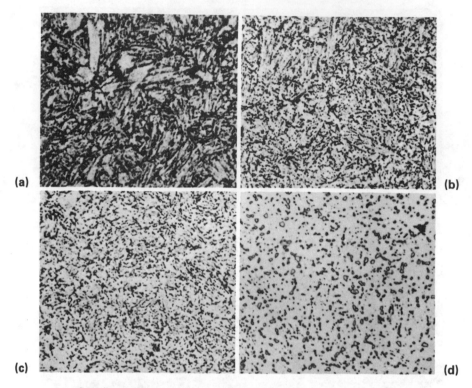

(a)

(b)

(c)

(d)

Samples were austenitized at 850 °C (1560 °F), water quenched, and tempered at 600 °C (1110 °F) for (a) 5 min, (b) 1 h, (c) 4 h, and (d) 32 h. Picral etch. Magnification, 1000×.

Fig. B.10. Tempered martensite microstructures in 1040 steel (Ref 132)

The tubing was normalized by austenitizing at 845 °C (1550 °F) and air cooling, and was surface hardened and tempered by induction heating. Top (surface layer) is tempered martensite; bottom (interior region) exhibits a pearlite matrix with grain-boundary ferrite. Nital etch. Magnification, 100×.

Fig. B.11. Near-surface region of normalized and surface hardened and tempered 1050 steel tubing (Ref 133)

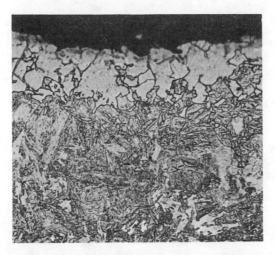

Decarburized (ferrite) surface layer at top; tempered martensite core below. Nital etch. Magnification, 500×.

Fig. B.12. Microstructure of hardened and tempered 10B35 steel bar (Ref 134)

was not austenitized at the hardening temperature employed (see Fig. 2.3); even if it had been, the hardenability of low-carbon austenite is very low. Decarburization may occur *prior to* (for example, during preheating prior to hot rolling) as well as *during* heat treating. In either case, its effects on hardening and tempering, particularly in surface heat treating, should be carefully considered.

Appendix C

Fundamentals of Electricity

A thorough understanding of induction heating requires a utilization of various principles of electricity and magnetism. Although an in-depth discussion of these concepts is beyond the scope of this book, a brief review of the important relations in elementary electricity is presented as a refresher for those who have had an introductory or an advanced course in this area. The interested reader is referred to college physics and electrical engineering texts for a more complete treatment (Ref 39, 41, and 135).

NOMENCLATURE

The discussion below makes frequent use of a number of quantities. Abbreviations and units of these quantities are as follows:

t ≡ time (seconds)
q ≡ electric charge (coulombs)
I ≡ current = dq/dt (amperes)
E ≡ electromotive force (EMF) (volts)
V ≡ potential drop (volts)
R ≡ resistance (ohms)
C ≡ capacitance (farads)
L ≡ inductance (henrys)
Z ≡ impedance (ohms)
X_L ≡ inductive reactance (ohms)
X_C ≡ capacitive reactance (ohms)
f ≡ frequency (hertz, or cycles per second)
ω ≡ angular frequency (radians per second)
$\mathbf{B}$ ≡ magnetic induction field vector (weber/meter^2 or tesla)

Following the more widely accepted convention (first proposed by Benjamin Franklin), electric currents will be taken to be from positive to negative—i.e., the current follows the direction of positive charge carriers—even though in reality electric currents are known to consist of electromagnetic waves propagated in the opposite sense by negative charge carriers (electrons).

DIRECT-CURRENT CIRCUITS

Direct-current (dc) circuits consist of complete electrical paths in which the current flows in a fixed direction. Typically, they consist of a source of electromotive force E, such as a battery, and a number of other components such as resistors, capacitors, and inductors (coils). The analysis of dc circuits makes frequent use of two principles known as Kirchoff's Laws, which are as follows:

1. In any circuit the total amount of current leaving any point must be equal to the amount approaching that point.
2. In any circuit, the total voltage drop around any complete path is exactly equal to zero. An alternative way of expressing this fact is that the voltage difference between two points is the same irrespective of the path taken in going from one point to the other.

The application of these principles will be illustrated for circuits containing the various types of electrical elements mentioned above.

DC Circuits With Batteries and Resistors

The simplest dc circuits contain only batteries and resistors. Resistors are circuit elements across which the potential (or voltage) *drop* V is directly proportional to the current I. The constant of proportionality is the resistance R, and the mathematical relationship expressing this linear dependence is known as Ohm's Law:

$$V = IR \qquad\qquad\qquad\qquad \text{(Eq C.1)}$$

In a dc circuit containing only a battery supplying an EMF of voltage E and a resistor of resistance R, Kirchoff's Second Law and Ohm's Law lead to the relation $E + (-IR) = 0$ or $I = E/R$. Note that the EMF represents an *increase* in voltage and the resistance a *decrease* or drop in voltage.

Kirchoff's Laws can also be used to analyze circuits with more than one resistor. For the case of resistors in series (Fig. C.1a), Kirchoff's First Law leads to the conclusion that the current through each resistor is equal. Then, Kirchoff's Second Law may be applied to the circuit to obtain $E - IR_1 - IR_2 - IR_3 = 0$, or $I = E/(R_1 + R_2 + R_3)$. In other words, the resistance equivalent to a *series* of resistors is one whose magnitude is simply equal to the sum of the individual resistances:

$$R_{series} = R_1 + R_2 + R_3 + \dots \qquad\qquad\qquad \text{(Eq C.2)}$$

For the case of resistors in parallel (Fig. C.1b), Kirchoff's Second Law demonstrates that the voltage drop across each resistor is equal to the EMF E; thus, a different current (E/R_i) flows through each resistor R_i. Kirchoff's First Law gives rise to the fact that the total current I_0 is equal to $I_1 + I_2 + I_3$. Therefore, if the three resistors in Fig. C.1(b) were replaced by a single resistor R_0 through which

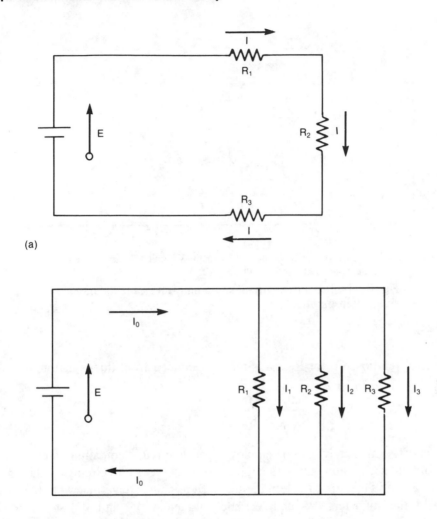

(a)

(b)

Fig. C.1. Direct-current circuits consisting of (a) a battery and three resistors in series and (b) a battery and three resistors in parallel

a current I_0 flowed, its value would be given by:

$$\frac{E}{R_0} = \frac{E}{R_1} + \frac{E}{R_2} + \frac{E}{R_3}$$

or

$$\frac{1}{R_0} = \frac{1}{R_1} + \frac{1}{R_2} + \frac{1}{R_3}$$

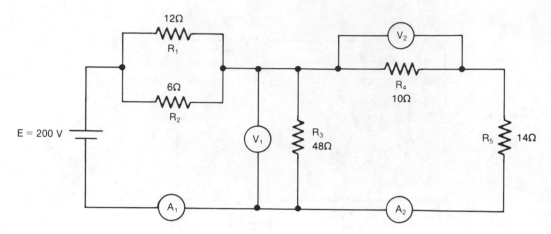

Voltage-drop readings are indicated by V1 and V2, and current readings are indicated by A1 and A2.

Fig. C.2. Direct-current circuit containing several resistors in series and in parallel

A general relationship for $R_{parallel}$ is readily deduced from this equation:

$$\frac{1}{R_{parallel}} = \frac{1}{R_1} + \frac{1}{R_2} + \frac{1}{R_3} + \ldots \qquad \text{(Eq C.3)}$$

Kirchoff's Laws may also be employed for dc circuits containing a number of resistors in series and parallel combinations. One such circuit is shown in Fig. C.2. The object is to determine the current and voltage that would be indicated by the ammeters (A1 and A2) and voltmeters (V1 and V2) shown in the figure. This is accomplished through the following calculations:

- Equivalent resistance for R_1 and R_2: $\left(\frac{1}{12} + \frac{1}{6}\right)^{-1} = 4$ ohms
- Equivalent resistance for R_3, R_4, R_5: $\left(\frac{1}{48} + \frac{1}{10 + 14}\right)^{-1} = 16$ ohms
- A1 = 200/(16 + 4) = 10 amps
- Voltage drop across R_1, R_2 = 10 × 4 = 40 volts
- Voltage drop across R_3 = 200 − 40 = 160 volts
- Voltage drop across R_4, R_5 = 160 volts
- A2 = 160/24 = 6.67 amps
- V2 = 6.67 × 10 = 66.7 volts.

In more complicated circuits, Kirchoff's Second Law is used to derive a set of simultaneous linear equations in terms of the currents in each branch of the circuit. One equation is developed for each complete circuit, and the set is solved by standard techniques.

DC Circuits With Resistors and Capacitors

A capacitor is a circuit element consisting of two conductors, or plates, on which equal and opposite charges may be established by connecting it to a source of EMF. The capacitance C is equal to the quotient of the charge q on either plate and the voltage difference between the plates:

$$C = \frac{q}{V}$$ (Eq C.4)

Note that C is a fixed quantity which depends on the geometry of each plate, the spatial relationship between them, and the medium ("dielectric") between them.

Relationships for capacitors in series and parallel are easily derived. These are:

$$C_{series} = \left(\frac{1}{C_1} + \frac{1}{C_2} + \frac{1}{C_3} + \ldots\right)^{-1}$$ (Eq C.5)

and

$$C_{parallel} = C_1 + C_2 + C_3 + \ldots$$ (Eq C.6)

Equation C.5 results from the fact that the *charge* on capacitors in series must be equal. Setting this charge equal to q, the net voltage drop across the series is equal to $V = V_1 + V_2 + V_3 = q/C_1 + q/C_2 + q/C_3 + \ldots$, which after rearrangement yields $C_{series} = q/V$ as defined above. Similarly, Eq C.6 is derived from the fact that the voltage drop across capacitors in parallel is constant. The net charge on the capacitors is then the sum of the individual charges $q = q_1 + q_2 + q_3 + \ldots = (C_1 + C_2 + C_3 + \ldots)V$, leading to the above equation for $C_{parallel}$.

When a capacitor is used as a dc circuit element, the current through the circuit is not constant. It decreases from some peak value while the capacitor is being charged by the source of EMF, and it increases when the capacitor is discharged. The analysis of a dc circuit containing a battery, a capacitor, and a resistor is straightforward and can be carried out using Kirchoff's Laws in much the same manner as for circuits containing only a battery and resistors.

The simplest "RC" circuit is shown in Fig. C.3(a). When the switch S is thrown, a closed circuit is produced. Applying Kirchoff's Second Law gives:

$$E - IR - (q/C) = 0$$ (Eq C.7)

Because $I = dq/dt$, this relationship can be transformed into the following linear differential equation for the charge on the capacitor:

$$R\frac{dq}{dt} + \frac{q}{C} = E$$ (Eq C.8)

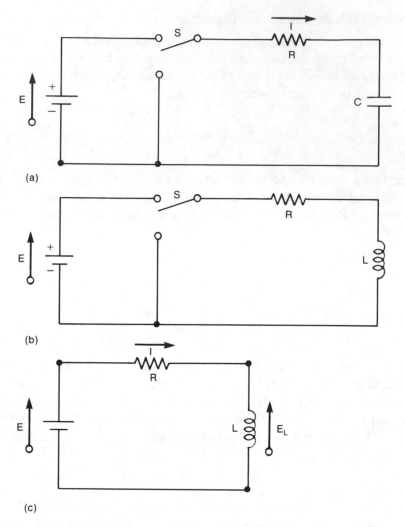

The illustration in part (c) indicates the direction of the EMF
induced by the inductor when the switch in the circuit shown in (b)
is closed.

**Fig. C.3. Direct-current circuits containing a battery, a resistor,
and (a) a capacitor or (b, c) an inductor (Ref 41)**

Although this equation is not difficult to solve, for the sake of brevity only the
solution will be presented:

$$q = CE[1 - \exp(-t/RC)] \qquad \text{(Eq C.9)}$$

The time derivative of Eq C.9 yields the current through the capacitor (and the
resistor):

$$\frac{dq}{dt} = I = \frac{E}{R} \exp(-t/RC) \qquad \text{(Eq C.10)}$$

Equation C.9 demonstrates that the charge on the capacitor builds up exponentially until it "saturates" at a value of CE at long times. Simultaneously, the current through the circuit (Eq C.10) *decreases* exponentially. Both of these processes are characterized by the capacitive time constant, RC, or the time at which the charge on the capacitor has increased to within a factor of $(1 - e^{-1}) = 0.63$ of its equilibrium value, CE. The simple circuit of Fig. C.3(a) and the above solution also establish that the peak current occurs *before* the peak voltage ($= q_{max}/C$) across the capacitor; the current is said to *lead* the voltage. In an ac circuit, the current across a capacitor leads the voltage by one-fourth of the ac cycle, or 90°.

DC Circuits With Resistors and Inductors

As its name implies, an inductor is a circuit element in which an EMF is *induced* by a *change* in magnetic flux through it. The simplest inductors are helical coils of conducting wire. The induced EMF may be generated by a change in flux generated by an adjacent coil or by a change in flux resulting from a change in the current in the original coil itself. In the latter instance, the EMF is self-induced. In either case, the induced EMF, E_L, is found from Faraday's Law:

$$E_L = \frac{d(N\Phi_B)}{dt} \qquad \text{(Eq C.11)}$$

Here, N is the number of turns in the coil and Φ_B is the magnetic flux set up in each turn. For coils in which no magnetic materials are nearby, $N\Phi_B = LI$, in which L is known as the inductance of the device. The induced EMF can then be written as:

$$E_L = -L\frac{dI}{dt} \qquad \text{(Eq C.12)}$$

When Eq C.12 is rearranged to express L in terms of E_L and dI/dt, it serves as the defining equation for the inductance of a coil of arbitrary shape and size.

The minus sign in Eq C.12 is very important: it establishes the fact that the induced EMF opposes an increase in current. A more general statement of this fact is given in Lenz's Law: "The induced voltage (and current) appears in such a direction that it opposes the change that produced it." Thus, if the current tends to *decrease* through an inductor, a self-induced EMF will be set up which will generate a current parallel to the original current.

Analysis of dc circuits with inductors involves the solution of linear differential equations which are similar to the one discussed above for the RC circuit. A simple "LR" circuit is shown in Fig. C.3(b). When the switch S is thrown, the situation in Fig. C.3(c) applies, and Kirchoff's Second Law may be utilized. The battery produces an increase in voltage equal to E, and a voltage drop of $-IR$ occurs across the resistor. Furthermore, the induced EMF, E_L, generated by the inductor opposes the increasing current as shown. Therefore, it is analogous to a voltage *drop* whose magnitude is equal to $-L(dI/dt)$. Summing the three voltages results in the following differential equation for I:

$$E - IR - L\frac{dI}{dt} = 0$$

or

$$L\frac{dI}{dt} + IR = E \qquad \text{(Eq C.13)}$$

The solution of Eq C.13 is:

$$I = \frac{E}{R}[1 - \exp(-Rt/L)] \qquad \text{(Eq C.14)}$$

In contrast to the current-versus-time behavior in an RC circuit, the current *increases* in an LR circuit when the switch is closed. At long times, it achieves an equilibrium value of E/R, or the current that would flow in the circuit if only the resistor were present. Differentiation of Eq C.14 leads to an expression for the induced EMF across the inductor, $E_L = -L(dI/dt)$:

$$E_L = -E \exp(-Rt/L) \qquad \text{(Eq C.15)}$$

At time $t = 0$, $E_L = -E$, and no voltage drop or current is developed across the resistor, as could have been deduced from Eq C.14. Also, it can be seen that the *magnitude* of E_L decreases exponentially with time. For this reason, the voltage across an inductor is said to *lead* the current. In an ac circuit, the voltage leads the current by one-fourth of the ac cycle, or 90°, for a pure inductor.

Equation C.14 can be employed to derive an inductive time constant, L/R, or the time at which the current in the circuit will reach a value within $1/e$ (~0.37) of its final equilibrium value.

ALTERNATING-CURRENT CIRCUITS

Alternating current (ac) circuits consist of complete electrical paths in which the current flows in alternating (i.e., reversing) directions. They usually consist of a source of EMF of some frequency f (in hertz, or cycles per second) and various resistors, capacitors, inductors, and possibly other electrical components. The analysis of such circuits also makes use of Kirchoff's Laws and will be discussed below.

Energy Storage and Dissipation in Electric Circuits

Electrical energy may be stored or dissipated by various electrical elements, be they in dc or ac circuits. An understanding of energy partitioning is especially useful in the analysis of ac circuits, including those used in induction heating applications.

In a resistor, energy is dissipated as Joule heating. This occurs at a rate (or power) equal to $-I^2R$ or $-VR$, the minus sign indicating that energy is being lost. In contrast to resistors, capacitors and inductors *store* energy in either an electric field (capacitor) or a magnetic field (inductor). For a capacitor, the stored energy is $U_E = q^2/2C$, whereas for an inductor, it is equal to $U_B = LI^2/2$.

The dissipation and storage of energy has an analogy in mechanical systems consisting of a spring, a mass, and a dashpot. Resistors are similar to dashpots. In addition, capacitors store potential energy in much the same way that springs do, and the kinetic energy associated with the motion of the mass m may be associated with the energy stored in an inductor. In fact, if the spring constant, displacement, and velocity are denoted by k, x, and v, the following correspondence is found between electrical and mechanical systems:

> q corresponds to x
> I corresponds to v
> C corresponds to 1/k
> L corresponds to m.

The energy stored by a spring is $\frac{1}{2}(kx^2)$, and the kinetic energy of the mass is $\frac{1}{2}(mv^2)$. Using the above analogs, the similarity between the capacitor and the spring $[U_E = \frac{1}{2}(q^2/C)$ versus $\frac{1}{2}(kx^2)]$ and that between the inductor and mass $[U_B = \frac{1}{2}(LI^2)$ versus $\frac{1}{2}(mv^2)]$ are apparent. Because of this analogy, the variations of current and voltage in ac circuits are commonly referred to as electromagnetic *oscillations*.

In an electrical system containing no resistance, free oscillations in current and voltage (i.e., those not driven by a source of EMF) will continue indefinitely. The frequency of such oscillations can be inferred from those of a spring-mass system, for which the natural frequency in hertz is given by $f = \sqrt{k/m}/2\pi$. Replacing k with $1/C$ and m with L leads to the natural frequency of an LC circuit, $f = (\sqrt{1/LC})/2\pi$. A similar LC circuit containing a small amount of resistance, known as a parallel resonant, or tank, circuit, is discussed later in this appendix.

Forced Oscillations and Resonance-Series LCR Circuits

Perhaps the most common and fundamental of all ac circuits is the LCR series circuit containing a source of alternating EMF and an inductor, a capacitor, and a resistor in series (Fig. C.4). If the ac voltage is assumed to be of the form $E = E_m \cos \omega t$, where ω is the (fixed) angular frequency (in radians per second), summation of the voltages around the LCR circuit yields the expression:

$$L\frac{dI}{dt} + RI + \frac{1}{C}q = E_m \cos \omega t \qquad \text{(Eq C.16)}$$

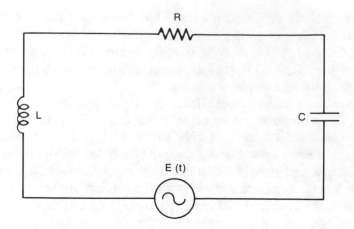

Fig. C.4. Simple LCR (inductor-capacitor-resistor) series alternating-current electrical circuit (Ref 41)

Replacing I with dq/dt leads to the second-order linear differential equation for q(t):

$$L\frac{d^2q}{dt^2} + R\frac{dq}{dt} + \frac{q}{C} = E_m \cos \omega t \qquad \text{(Eq C.17)}$$

The similarity of this equation with the equation of motion for a spring-mass-dashpot system driven by a periodic external force of magnitude F_m and angular frequency ω is evident:

$$m\frac{d^2x}{dt^2} + b\frac{dx}{dt} + kx = F_m \cos \omega t \qquad \text{(Eq C.18)}$$

The solution of Eq C.17 is:

$$q = \frac{E_m}{G} \sin (\omega t - \phi) \qquad \text{(Eq C.19)}$$

where

$$G = \sqrt{\left(\omega^2 L - \frac{1}{C}\right)^2 + R^2\omega^2}$$

and where ϕ, the "phase angle" between the "driving force" (the EMF) and the response, is given by:

$$\phi = \cos^{-1}\left(\frac{R\omega}{G}\right)$$

Differentiation of Eq C.19 yields an expression for the current, which is the *same* through each of the series circuit elements:

$$I = \frac{\omega E_m}{G} \cos(\omega t - \phi) = I_m \cos(\omega t - \phi) \qquad \text{(Eq C.20)}$$

where

$$I_m = \frac{\omega E_m}{G} = E_m \Big/ \sqrt{\left(\omega L - \frac{1}{\omega C}\right)^2 + R^2}$$

Inspection of Eq C.20 establishes that the current will have its maximum amplitude when $\omega L = 1/\omega C$ or when $\omega = \sqrt{1/LC}$, the natural frequency of the LC combination. This frequency is also called the resonant frequency of the LCR circuit. At resonance, $I_m = E_m/R$, and the circuit appears to be a pure resistance.

Often, expressions such as Eq C.20 are plotted to show the "sharpness" of the relationship between I and ω. Several examples, or "resonance curves," are shown in Fig. C.5. The sharpness of such curves is measured by their "half-widths." The half-width $\Delta\omega$ is the difference between two frequencies, each of which corresponds to a current amplitude of one-half of the maximum current amplitude. Note

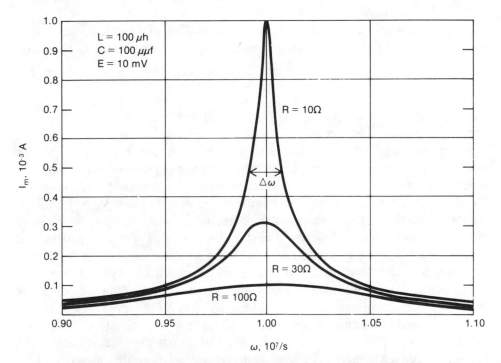

Fig. C.5. Current amplitude as a function of frequency for an LCR series ac circuit whose components have the values indicated (Ref 41)

that if the resistance is large, the curve is broader and flatter than if it is small. Thus, small-R circuits are better able to discriminate against frequencies on either side of resonance. This property is often used to pass current of a certain frequency while limiting the flow of current at other frequencies.

Reactance and Impedance

The above analysis of the LCR series circuit reveals several important quantities used frequently in the description of ac circuits found in induction heating installations. The first two of these quantities are the inductive reactance, X_L, and capacitive reactance, X_C. These quantities (which are measures of the opposition to the flow of alternating current presented by inductors and capacitors) appear in the square root denominator defining I_m in the LCR circuit (Eq C.20). They are defined as:

$$X_L = \omega L = 2\pi fL \qquad \text{(Eq C.21a)}$$

and

$$X_C = 1/\omega C = 1/2\pi fC \qquad \text{(Eq C.21b)}$$

in which the relationship between cyclic frequency f and angular frequency ω ($f = \omega/2\pi$) has been employed.

Another important quantity is the over-all impedance of the circuit, denoted by Z. It is defined as the square root denominator in the expression for I_m above, or:

$$Z = \sqrt{(X_L - X_C)^2 + R^2} \qquad \text{(Eq C.22)}$$

Note that at resonance, $X_L = X_C$, $Z = R$, and the impedance to current flow appears as a pure resistance.

A physical explanation for the form of the equation for Z can be gained by analyzing the variation of current and voltage through inductors, capacitors, and resistors in an ac circuit in which the voltage varies in a sinusoidal fashion. Using the simple relations $E_L = -V_L = -L(dI/dt)$, $C = q/V_C$, and $V_R = IR$ for inductors, capacitors, and resistors, the phase, or angular, relationship between current and voltage in each device is readily determined. For a pure inductor, the current sinusoid *lags* the voltage sinusoid by one-fourth of a cycle (90°), whereas, for a capacitor, the current *leads* the voltage by one-fourth of a cycle. These phase relationships are represented pictorially through the use of vectors, as shown in Fig. C.6. The net voltage drop is, therefore, equal to $\sqrt{(V_L - V_C)^2 + V_R^2}$, which, of course, must be equal to the applied EMF, $E = E_m \cos \omega t$. By using the definitions of V_L, V_C, and V_R, it is readily found that the voltages across the various circuit elements, disregarding the phase relationship, are $V_L = IX_L$,

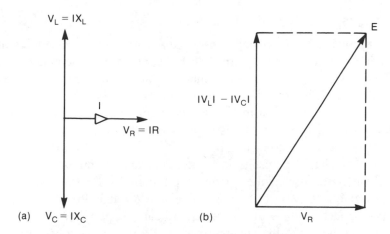

(a) Diagram illustrating the phase relationships between current and voltage in the various components of an LCR series ac circuit.
(b) Diagram illustrating the relationship between the applied EMF E and the voltage drops across the resistor (V_R), the inductor (V_L), and the capacitor (V_C) in such a circuit.

Fig. C.6. Vector diagrams for an LCR circuit (Ref 46)

$V_C = IX_C$, and $V_R = IR$. Summing these quantities vectorially, as is done in Fig. C.6, in order to restore the phase relationship, leads to the following:

$$\text{Applied EMF} = I\sqrt{(X_L - X_C)^2 + R^2} = IZ \qquad \text{(Eq C.23)}$$

Equation C.23, relating the EMF to the current, with the proportionality constant being the impedance, is the ac circuit equivalent of Ohm's Law.

The definition of the inductive reactance allows the determination of an alternative means of quantifying the sharpness of resonance of a tuned LCR circuit. This is accomplished through the Q factor:

$$Q = X_L/R = 2\pi fL/R \qquad \text{(Eq C.24)}$$

The values of Q for a resonant circuit typically range between 20 and 100. The Q factor also offers a comparison between the total energy in a tuned circuit and the energy which is dissipated by the resistance in a circuit. Since no energy is dissipated in a pure inductance or in a pure capacitance, the resistance should be kept as low as possible in order to reduce energy loss. In induction heating, because nearly all of the resistance of a resonant circuit is in the coil, the induction coil resistance should be kept low for maximum efficiency. In these instances, efficiency is equal to the ratio of the power dissipated in the load to the power dissipated in the load *and* the heating coil.

True, Reactive, and Apparent Power

The definitions of reactance and impedance allow several other important quantities in induction heating to be defined. These are the true, reactive, and apparent power. True power, P_T, is the power which is available to the coil and workpiece for actual heating. It is expressed in kilowatts (kW) and is equal to I^2R. Apparent power, P_A, is equal to I^2Z, or total volts times total amperes. To differentiate it from true power, apparent power is expressed in kilovolt-amperes (kVA). The third power quantity is the reactive power, $P_R = I^2(X_L - X_C)$; this is the power which is out of phase. It results in no heating, and its units are expressed as kilovars (kvar). Because of the relationship among Z, R, and $(X_L - X_C)$, the various power quantities are themselves related through $P_A = \sqrt{P_R^2 + P_T^2}$.

The "power factor" for a given induction heating operation is defined as the ratio of the true power to the apparent power, P_T/P_A. In induction heating, it is desirable to have a power factor at the power source as close to unity as possible.

The power factor may also be defined as the cosine of the phase angle between the current and voltage in an electric circuit. For a tuned series LCR circuit which appears to consist of a pure resistance, the current and voltage are in phase, $P_R = 0$, and the power factor is equal to unity. For both a pure inductor and a pure inductance, the phase angle is equal to 90°, $P_T = 0$, and the power factor viewed across these elements alone is equal to zero. Note also that low power factors correspond to high values of Q.

Parallel Resonant Circuits

Probably the most common ac circuit used in induction heating systems is the parallel resonant circuit. The simplest type consists of an inductor in parallel with a capacitor; associated with the inductor is a resistance. This type of circuit (Fig. C.7) is called a "tank circuit." The circuit serves as a "storage tank" for

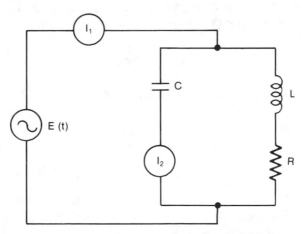

Fig. C.7. Simple parallel resonant (tank) circuit consisting of a supply voltage E and a capacitor in parallel with an inductor and a resistor

electromagnetic energy. The impedance of a parallel resonant circuit is equal to $(2\pi fL)^2/R$ or $(2\pi fL)Q$. At resonance, Q is high and thus the impedance is high.

The parallel resonant (tank) circuit shown in Fig. C.7 is tuned by making the inductive and capacitive reactances equal. When this is done, the currents through the two are equal and opposite in phase. Hence, they cancel each other in the external line circuit, and the line current I_1 is thus very small. The small line current results from the fact that the inductor has a small but finite resistance causing a slight phase angle shift so that complete current cancellation cannot take place. On the other hand, the circulating current shown as I_2 is very large, depending upon the applied voltage and the reactance of the capacitor at the resonant frequency $(= 1/2\pi\sqrt{LC}$, as in a resonant series LCR circuit). In a parallel resonant, or tank, circuit, therefore, the impedance is a maximum across the L-C circuit, and the line current is very small. Also, the power factor as viewed across the voltage terminals is near unity.

Transformers and Impedance Matching

Matching of induction heated loads to induction generators often makes use of transformers. A transformer is a device consisting of two coils (or windings) which have "mutual" inductance between them. The *primary* winding is connected to the ac supply, and a voltage is induced in the *secondary* winding, which is separated from the primary by an iron core or air core. A transformer can be used to increase ("step up") or decrease ("step down") voltages.

The relationship between the voltages in the primary (E_P) and secondary (E_S) is determined by the ratio of the turns in each:

$$\frac{E_P}{E_S} = \frac{N_P}{N_S} \qquad \text{(Eq C.25)}$$

Here, N_P and N_S denote the turns in the primary and secondary, respectively. The current that flows in the secondary winding as a result of the induced voltage must produce a flux which exactly equals the primary flux. Since the flux is proportional to the product of the number of turns and the currents in the windings, the relationship between the primary current I_P and the secondary current I_S is:

$$N_P I_P = N_S I_S$$

or

$$\frac{N_P}{N_S} = \frac{I_S}{I_P} \qquad \text{(Eq C.26)}$$

Hence, when the voltage is stepped up, the current is stepped down, and vice versa. It should be borne in mind, however, that Eq C.25 and C.26 apply only to transformers with perfect coupling between the primary and secondary windings.

In reality, coupling is not perfect, and the voltage and current ratios will each be slightly less than the turns ratio.

Induction heating sources have rated current and voltage limits that cannot be exceeded without damage to the source. The ratio of the rated voltage to the rated current is the effective impedance of the source. To obtain the greatest transfer of energy from the source to the load, the impedances of the two should be as close as possible to each other. If they do not match, transformers are then employed: the impedance of the secondary is adjusted to match that of the electrical load, and the primary impedance matches that of the source. Usually this involves reducing a high-voltage line supply to a lower-voltage, higher-current one. Because impedance is equal to E/I, the impedance ratio between the primary and secondary (Z_P/Z_S) can be found from Eq C.25 and C.26:

$$\frac{Z_P}{Z_S} = \frac{N_P^2}{N_S^2} \qquad\qquad \text{(Eq C.27)}$$

Thus, the impedance ratio is equal to the *square* of the turns ratio.

IMPEDANCE MATCHING AND TUNING OF INDUCTION HEATING SYSTEMS

Fixed-Frequency Sources

Impedance matching and tuning of fixed-frequency induction systems both make use of the principles described above. Impedance matching in induction heating (and other) systems enables the rated amount of power to be drawn from the generator. The actual power drawn for a given coil is equal to the I^2Z, where Z is the unity-power-factor impedance of the coil and workpiece reflected in the coil circuit. Since $I = E_g/Z$, the power is also equal to E_g^2/Z, where E_g is the generator output voltage. If the Z of the coil is not essentially equal to that of the generator, it must be adjusted for purposes of matching. This is typically done using a transformer as discussed above.

In addition to impedance matching, the induction coil circuit must be tuned to resonance for maximum power transfer. This is done using a capacitor in parallel with the induction coil. In fact, the circuit must be tuned before the proper transformer ratio is calculated as described above, because the calculation of the transformer ratio is more easily made from the unity-power-factor voltage, current, and impedance. The unity power factor is achieved by balancing the reactance of the coil and the reactance of the capacitor. Once the proper capacitance is selected, the parallel combination of induction heating coil and capacitor draws essentially the same current from the power supply as if it were a pure resistance. Although the current drawn from the generator is low, being only enough to supply the heating power, high currents circulate between the coil and the capacitor. A computer program for calculating coil turns and values of capacitance for load matching to a fixed-frequency source is given in Appendix D.

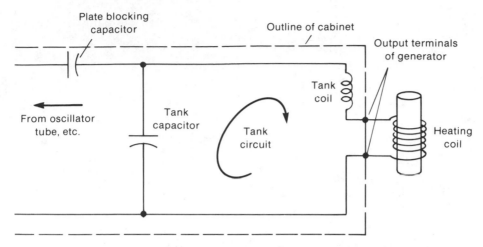

Fig. C.8. Typical setup of a tank circuit used in induction heating employing an RF (vacuum tube) power supply (Ref 46)

RF Sources

In induction heating applications involving RF power supplies, the heating coil is usually connected in series with another, usually larger inductor element in the appropriate leg of a parallel resonant (tank) circuit (Fig. C.8). In this type of setup, the tank current, rather than the voltage across the induction coil, tends to be constant.

The RF generator consists of a power supply, a vacuum tube, and a tank circuit. The vacuum tube can supply energy at a range of frequencies. The specific frequency is determined by the parameters of the tank circuit. Often, the tap settings on the internal tank coil must be adjusted to provide a resonant frequency within the range of the vacuum tube.

Induction coils for RF heating applications are usually matched to the power supply by choosing the number of turns. Sometimes output transformers and modifications to the tank circuit inductor are also utilized. However, because the loads in RF situations tend to be small, coil matching is usually not extremely critical.

Appendix D

Design of Induction Coils

One of the most important considerations in setting up an induction heating system is the problem of designing the coil for the specific workpiece (load) to be heated. As discussed in Chapter 4 and Appendix C, it is necessary to match and tune the induction coil (with its enclosed load) to the power source in order to obtain the desired power output from that source. Essentially, the procedure involves designing a coil with a specific number of turns and determining the proper capacitance to tune the load coil tank circuit to resonance (that is, to obtain a unity power factor). Associated electrical parameters are often desired from the design as well. These include the current in the tuned circuit, the power lost in the induction coil and the transmission line, and the efficiency of the system.

Coil design is usually more critical when using low-frequency supplies such as motor-generator sets or solid-state power supplies which provide essentially a fixed frequency and fixed impedance. It is seldom critical in situations employing RF frequencies because the load is usually small, and, therefore, the power requirement is low and the system is self-resonating. However, it is sometimes desirable to get an approximate induction coil design even for RF systems to obtain optimum results.

Induction coil parameters can be calculated very accurately for a *cylindrical* load. Such calculations have been described several times in the literature. The calculation method described in this appendix is based on that developed by Vaughan and Williamson (Ref 52 and 53); coil designs based on this technique have been extensively verified using experimental methods. The calculation is accurate within 10% for long, closely coupled solenoid coils. It is still usable with possibly 25% accuracy if the coil is short and loosely coupled (for example, coil-length-to-coil-diameter ratio between 0.5 and 1 and load-diameter-to-coil-diameter ratio of 0.3 to 0.5).

DESIGN PROCEDURE/COMPUTER PROGRAM

To calculate the coil turns and associated electrical parameters for a given load, the following information is needed:

1. A generator frequency must be specified. Considerations for frequency selection are described in Chapter 3. In essence, the frequency is chosen so that the diameter of the load cylinder is more than four times the reference or skin depth (D2). The frequency must sometimes be that of a generator that is available on site but that may not necessarily be the best choice. However, assuming that there is a wide choice, a generator of the desired frequency and power output level, with an input power and voltage that match the available line power, is used. This provides the output power, voltage, and current for the calculations discussed below.

2. The over-all geometry of the coil and load must be established (Fig. D.1). These parameters are often constrained by the situation such as the need for thermal insulation or a protective atmosphere enclosure or other space limitations. In general, the coil should be of about the same length as that of the load, the coupling should be as close as possible, and the space factor (the ratio of coil-turns width to total coil length—that is, the amount of space filled by the coil) should be 0.8 or 0.9.* The generator frequency, the generator electrical parameters, and the load geometry provide all the information necessary to calculate the desired number of coil turns and the tank-circuit capacitance. To permit a better understanding of the items in the computer program below, Fig. D.2 shows the equivalent coil circuit, including a resonating capacitor and the transmission line between the coil and the capacitor. The individual components are referred to also in Table D.1, which lists the input variables for the computer calculation outlined below.

Most of the input data for the computer program (Table D.1) are self-explanatory. However, some items require a brief explanation:

- If the load is magnetic, an estimated power PES must be entered to permit calculation of load resistance. Usually about 90% of PL is used for this value.
- If the load is not magnetic, a value of skin or penetration depth D2 must be calculated as discussed in Chapter 3:

$$D2 = 3160(RHO2/F)^{0.5}$$

- If the load is a hollow cylinder, with the ratio of T2 (wall thickness) to penetration depth D2 between 0.5 and 2.0, values of K3 and K4 must be obtained

---

*Note that the (assumed) space factor and subsequently calculated number of coil turns establish the outer diameter of the inductor tubing; the calculation of required turns does not take the outer diameter of the coil tubing directly into account. If the number of required turns in the coil greatly exceeds that which can be physically wound in the available space, then a transformer must be used.

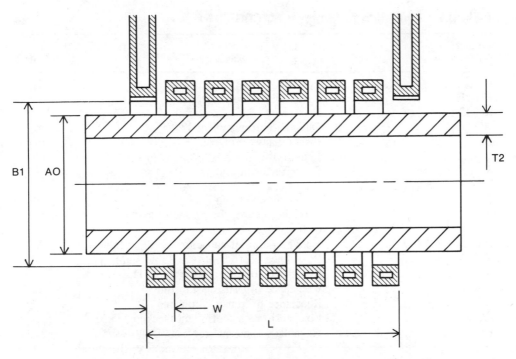

Fig. D.1. Schematic illustration of induction coil and load (Ref 52)

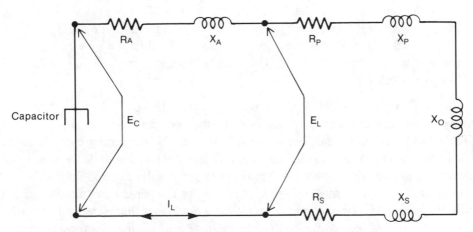

Fig. D.2. Equivalent electric circuit used in analysis for design of induction heating coils (Ref 52)

from Tables D.2 and D.3. If the workpiece is a hollow cylinder with a T2/D2 ratio less than 0.5 or greater than 2.0, K3 and K4 can be entered as zero.
- If the load is a solid cylinder, K2 must be obtained from Fig. D.3 by using the ratio of AO to D2.

Table D.1.   Input data for program FALCON

Variable designation	Variable	Units
F	Frequency	kHz
RHO1	Resistivity of coil material	Ω · in.
RHO2	Resistivity of load material	Ω · in.
AO	Outside diameter of load	in.
B1	Inside diameter of coil	in.
L	Length of coil	in.
S	Space factor of coil	. . .
T2	Wall thickness of load	in.
PL	Power into coil and load	kW
PES	Estimated power for magnetic load	kW
EL	Coil voltage	V
ER	Voltage rating of capacitor	V
K3	Resistance coefficient (see Table D.2)	. . .
K4	Reactance coefficient (see Table D.3)	. . .
RA	Resistance of transmission line	Ω
XA	Reactance of transmission line	Ω
K2	Load-resistance coefficient (see Fig. D.3)	. . .
K5	Nagaoka's constant (see Fig. D.4)	. . .

- For all loads, K5 must be obtained from Fig. D.4 by using the ratio of B1 to L.
- If the distance between the capacitor and coil is short — say within several feet — values of RA and XA usually are insignificant and can be neglected (i.e., taken equal to zero). However, in some cases, the distance may be large and these values must be calculated. They can be estimated from data in Davies and Simpson (Ref 136).

Figure D.5 gives a FORTRAN computer program (FALCON) that can be used to carry out the coil-design procedure. Output parameters from the program (Table D.4) include the number of turns in the coil N, the tuning capacitance in kVA, the current in the coil IL, and the efficiency ETA. It should be noted that capacitors in induction heating power supplies are usually rated in both kVA and microfarads but sometimes only in kVA. Many of the parameters are printed out only for reference. For example, Z, which is a parameter that is dependent on the geometry and frequency, can be used to evaluate quickly the relationship among turns, power, and voltage since $N = EL \cdot Z / \sqrt{PL}$.

SAMPLE TEST CASES

The examples described in Tables D.5, D.6, D.7, and D.8 give sample input and output data from program FALCON for several different cases. The different geometries are: a long, solid nonmagnetic cylinder; a long, solid magnetic cylinder; a short, thin-wall nonmagnetic cylinder; and a short, very-thin-wall

Table D.2. Resistance coefficients, K3, for hollow cylinders with ratios of load wall thickness to penetration depth (T2/D2) from 0.500 to 2.00 for various ratios of load wall thickness to load outside diameter (T2/AO) (Ref 52)

| T2/AO | Resistance coefficient(a), K3, for cylinder with T2/D2 ratio of:- | | | | | | | | | | | | | | | | |
	0.500	0.522	0.545	0.569	0.595	0.621	0.648	0.677	0.707	0.74	0.77	0.805	0.88	1.00	1.19	1.41	2.00
0........	2	1.93	1.86	1.78	1.70	1.64	1.58	1.51	1.45	1.39	1.34	1.29	1.20	1.08	0.97	0.92	0.94
0.025.....	1.96	1.89	1.83	1.76	1.70	1.64	1.58	1.51	1.45	1.39	1.34	1.29	1.21	1.10	0.99	0.93	0.95
0.0500....	1.77	1.75	1.72	1.68	1.64	1.58	1.54	1.49	1.44	1.39	1.34	1.30	1.23	1.11	1.04	0.95	0.96
0.0648....	1.60	1.60	1.59	1.58	1.56	1.53	1.50	1.46	1.42	1.37	1.33	1.29	1.23	1.12	1.04	0.95	0.96
0.077.....	1.45	1.47	1.48	1.48	1.48	1.47	1.45	1.43	1.39	1.36	1.32	1.29	1.23	1.13	1.04	0.96	0.96
0.084.....	(1.35)	1.38	1.40	1.42	1.44	1.43	1.42	1.40	1.37	1.34	1.32	1.28	1.22	1.13	1.04	0.97	0.96
0.092.....	(1.24)	(1.29)	1.32	1.35	1.37	1.38	1.38	1.36	1.35	1.33	1.31	1.27	1.21	1.13	1.04	0.97	0.96
0.100.....	(1.14)	(1.19)	(1.23)	1.27	1.30	1.32	1.34	1.34	1.34	1.32	1.30	1.27	1.21	1.13	1.04	0.98	0.97
0.109.....	(1.03)	(1.08)	(1.13)	(1.17)	1.21	1.24	1.26	1.27	1.28	1.27	1.27	1.25	1.20	1.13	1.05	0.98	0.97
0.119.....	(0.91)	(0.97)	(1.03)	(1.08)	(1.14)	1.17	1.20	1.22	1.24	1.24	1.24	1.23	1.19	1.13	1.05	0.99	0.97
0.130.....	(0.80)	(0.86)	(0.92)	(0.98)	(1.03)	(1.08)	1.12	1.16	1.18	1.19	1.20	1.20	1.18	1.13	1.06	0.99	0.98
0.141.....	(0.68)	(0.75)	(0.82)	(0.88)	(0.94)	(0.99)	(1.04)	1.08	1.12	1.14	1.16	1.17	1.17	1.13	1.06	1.00	0.98
0.154.....	(0.57)	(0.62)	(0.68)	(0.75)	(0.83)	(0.89)	(0.94)	(1.00)	1.04	1.08	1.11	1.13	1.16	1.13	1.07	1.01	0.98
0.168.....	(0.46)	(0.53)	(0.58)	(0.63)	(0.72)	(0.79)	(0.86)	(0.92)	(0.97)	1.02	1.06	1.09	1.14	1.13	1.08	1.01	0.99
0.183.....	(0.38)	(0.43)	(0.48)	(0.54)	(0.60)	(0.66)	(0.73)	(0.81)	(0.88)	(0.94)	1.00	1.04	1.09	1.13	1.09	1.02	0.99
0.200.....	(0.28)	(0.38)	(0.38)	(0.43)	(0.49)	(0.56)	(0.62)	(0.70)	(0.77)	(0.83)	(0.91)	0.97	1.06	1.12	1.10	1.03	1.00

(a) Values in parentheses not checked experimentally.

Table D.3. Reactance coefficients, K4, for hollow cylinders with ratios of load wall thickness to penetration depth (T2/D2) from 0.500 to 2.00 for various ratios of load wall thickness to load outside diameter (T2/AO) (Ref 52)

T2/AO	Reactance coefficient(a), K4, for cylinder with T2/D2 ratio of:															
	0.500	0.545	0.595	0.648	0.707	0.77	0.84	0.92	1.00	1.09	1.19	1.30	1.41	1.54	1.68	2.00
0.......	0.33	0.36	0.39	0.43	0.47	0.51	0.55	0.60	0.65	0.70	0.75	0.80	0.86	0.91	0.95	0.99
0.0063...	0.43	0.44	0.45	0.48	0.51	0.54	0.57	0.61	0.66	0.71	0.75	0.80	0.86	0.91	0.95	0.99
0.0088...	0.47	0.46	0.47	0.49	0.52	0.55	0.58	0.62	0.67	0.71	0.76	0.81	0.86	0.91	0.95	0.99
0.0125...	0.53	0.51	0.51	0.52	0.54	0.56	0.59	0.63	0.67	0.72	0.77	0.82	0.86	0.91	0.95	0.99
0.0177...	0.61	0.57	0.56	0.56	0.57	0.59	0.61	0.64	0.69	0.73	0.77	0.82	0.87	0.91	0.95	1.00
0.0210...	0.67	0.62	0.59	0.59	0.59	0.60	0.62	0.65	0.69	0.74	0.78	0.83	0.87	0.91	0.95	1.00
0.0250...	0.73	0.68	0.63	0.62	0.61	0.62	0.63	0.66	0.70	0.74	0.78	0.83	0.87	0.91	0.95	1.00
0.0297...	0.80	0.72	0.68	0.66	0.64	0.64	0.65	0.69	0.71	0.75	0.78	0.84	0.87	0.91	0.95	1.00
0.0354...	0.89	0.79	0.74	0.70	0.67	0.67	0.67	0.70	0.72	0.76	0.79	0.84	0.88	0.92	0.95	1.00
0.0420...	0.99	0.89	0.80	0.75	0.72	0.70	0.70	0.72	0.74	0.77	0.80	0.84	0.88	0.92	0.95	1.00
0.0500...	1.09	0.98	0.88	0.82	0.76	0.74	0.73	0.74	0.75	0.78	0.81	0.85	0.89	0.92	0.95	1.00
0.0595...	1.20	1.07	0.97	0.88	0.82	0.78	0.76	0.76	0.78	0.80	0.83	0.86	0.89	0.93	0.96	1.00
0.0707...	1.31	1.18	1.07	0.96	0.89	0.84	0.81	0.80	0.80	0.82	0.84	0.88	0.90	0.93	0.96	1.00
0.084...	(1.40)	1.30	1.19	1.07	0.97	0.90	0.86	0.84	0.84	0.84	0.86	0.89	0.91	0.93	0.96	1.00
0.100...	(1.46)	(1.38)	1.28	1.18	1.07	0.99	0.93	0.89	0.87	0.88	0.88	0.90	0.92	0.94	0.96	0.99
0.119...	(1.49)	(1.46)	(1.37)	1.27	1.17	1.08	1.01	0.96	0.92	0.91	0.90	0.92	0.93	0.95	0.97	0.99
0.141...	(1.44)	(1.46)	(1.43)	(1.36)	1.28	1.20	1.11	1.04	0.99	0.96	0.94	0.94	0.95	0.96	0.97	0.99
0.168...	(1.36)	(1.42)	(1.44)	(1.42)	(1.38)	1.30	1.22	1.14	1.07	1.02	0.98	0.97	0.97	0.97	0.98	0.99
0.200...	(1.23)	(1.34)	(1.40)	(1.44)	(1.44)	(1.41)	1.34	1.26	1.17	1.10	1.04	1.01	0.99	0.99	0.98	0.99

(a) Values in parentheses not checked experimentally.

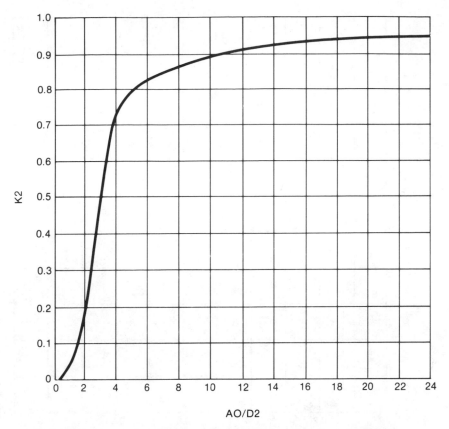

Fig. D.3. Load-resistance coefficient K2 as a function of the ratio of the outside diameter of the load to the reference, or skin, depth (Ref 52)

nonmagnetic cylinder. The specific dimensions can be found in the tables. Note that for the magnetic case (Table D.6), T2 has been set equal to zero even though the workpiece is a solid bar. The condition T2 = 0 is an artifact which makes the program solve the magnetic case. For *tubular* magnetic loads, it is assumed that the reference depth is much smaller than the wall thickness, thereby resulting in a solution equivalent to that which would be obtained for a *solid* magnetic workpiece.

These examples may be used as test cases by those implementing the program FALCON, or a similar code, on their own computing systems.

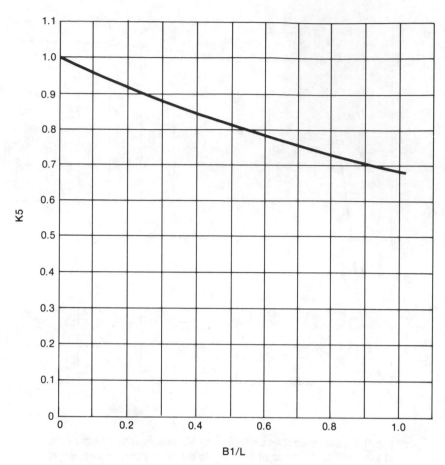

Fig. D.4. Nagaoka's constant K5 as a function of the ratio of the inside diameter of the coil to the coil length (Ref 52)

```
       PROGRAM FALCON(INPUT,OUTPUT,TAPE5=INPUT,TAPE6=OUTPUT)
C  PROGRAM TO CALCULATE  INDUCTION COIL TURNS AND CAPACITANCE FOR LOAD
C  MATCHING AND CIRCUIT TUNING IN INDUCTION HEATING
       REAL IL,K1,K1SQ,K2,K3,K4,K5,K7,K8,K9,KC,KL,KVA,L,N,NR,NSQ,NX
       PI = 3.1415927
       WRITE(6,8)
     8 FORMAT(1H1)
       WRITE(6,10)
    10 FORMAT(/////,
      +17X,          41H* * * * * * * * * * * * * * * * * * * * *, //,
      +17X,          41H*                 PROGRAM FALCON          *, //,
      +17X,          41H*          INDUCTION HEATING COILS         *, //,
      +17X,          41H*           FOR CYLINDRICAL LOADS          *, //,
      +17X,          41H* * * * * * * * * * * * * * * * * * * * *, //////)
C  ENTER INPUT DATA
       READ(5,20) F,RHO2,RHO1,AO
    20 FORMAT(4F15.9)
       READ(5,30) B1,L,S,T2
    30 FORMAT(4F15.9)
       READ(5,40) PL, PES,EL,ER
    40 FORMAT(4F15.9)
       READ(5,50) K3,K4,RA,XA
    50 FORMAT(4F15.9)
       READ(5,60) K5,K2
    60 FORMAT(2F15.9)
C  IF AOD IS LESS THAN OR EQUAL TO 2.0, ABORT AND INCREASE FREQUENCY
       FH = 1000.*F
       DO = 3160.*SQRT(1./FH)
       D2 = DO*SQRT(RHO2)
       B1OL = B1/L
       AOD = AO/D2
       IF(AOD.GT.2.0) GO TO 100
       WRITE(6,70) AOD
    70 FORMAT(//,
      +53H          PROGRAM ABORTING BECAUSE AO/D2 MUST BE GT 2.0, /,
      +46H          INCREASE FREQUENCY SINCE AO/D2 IS ONLY, F7.3, /)
       GO TO 9999
C  BEGIN CALCULATIONS
   100 D1 = DO*SQRT(RHO1)*SQRT(1./S)
       KC = PI*RHO1*B1/S
       KL = PI*RHO2*AOD
       AM = AO - T2
       AOB = AO/B1
       AOBSQ = AOB**2.
       K1 = K5*(1.-AOBSQ) + AOBSQ
       K1SQ = K1**2.
       K8 = 5.*PI*K1*1.E-08
       AOSQ = AO**2.
       B1SQ = B1**2.
       E = K8*FH*(B1SQ - AOSQ)/L
       C = KC/(D1*L)
       A = C+KC/(B1*L)
       TOA = T2/AO
       TOD = T2/D2
C  CHECK IF T2/D2 = 0.
       IF(TOD.EQ.0.) GO TO 200
C  CONTINUE FOR NONZERO VALUES OF T2/D2
```

Fig. D.5. Computer program FALCON used for design of induction heating coils (Ref 137)

```
           IF(TOD.GT.2.) GO TO 500
           IF(TOD.GT.0.5) GO TO 400
           IF(TOD.GT.0.) GO TO 300
C    SOLVE MAGNETIC CASE (T2/D2 = 0.)
       200 PAL = PI*AO*L
           ALGX = ALOG(PES) - 0.5*ALOG(FH)-ALOG(PAL)
           YY = ALOG(0.003) -  0.301*ALGX
           K7 = EXP(YY)
           PAOL = PI*AO/L
           B = PAOL*K1SO*SORT(FH)*K7*1.E-03
           D = 0.65*B
           GO TO 600
C    SOLVE CASE OF 0. LT T2/D2 LE 0.5
       300 AMSQ = AM**2.
           FAOL = (FH*AMSQ)/L
           K9 = K8*FAOL*K1
           D2SQ = D2**2.
           DOAT = 2.*D2SQ/(AM*T2)
           DOATSQ = DOAT**2.
           B = K9*DOAT/(1. + DOATSQ)
           AOMSQ = AOSQ/AMSQ
           D = K9*AOMSQ - K9/(1. + DOATSQ)
           GO TO 600
C    SOLVE CASE OF 0.5 LT T2/D2 LE 2.
       400 B = K1SQ*K3*KL*AM/(L*AO)
           D = K1SQ*K4*KL/L
           GO TO 600
C    SOLVE CASE OF T2/D2 GT 2.
       500 B = K1SQ*K2*KL/L
           D = B/K2
C    CALCULATE REMAINING VARIABLES
       600 AB = A+B
           ABSQ = AB**2.
           CDE = C+D+E
           CDESQ = CDE**2.
           AESQ = ABSQ + CDESQ
           SRAESQ = SORT(AESQ)
           Z = SORT(AB)/SRAESQ
           PLW = 1000.**L
           N = EL*Z/SORT(PLW)
           NSQ = N**2.
           NR = NSQ*AB+RA
           NX = NSQ*CDE+XA
           IL = EL/(NSQ*SRAESQ)
           EC = IL*SQRT((NR**2.)+(NX**2.))
           ETA = B/AB
           PS = ETA*PL
           PA = (IL**2.)*RA
           PT = (PLW + PA)/1000.
           KVA = PT*(ER**2.)*NX/((EC**2.)*NR)
C    PRINT INPUT / OUTPUT DATA
           WRITE(6,1100)
      1100 FORMAT(1X, 16HINPUT ==========, 3X, 1HF, 9X, 4HRHO1,
          +8X, 4HRHO2, 10X, 2HAO)
           WRITE(6,1110) F,RHO1,RHO2,AO
      1110 FORMAT(14X, 4F12.7, /)
           WRITE(6,1120)
```

Fig. D.5. (Continued)

```
 1120 FORMAT(20X, 2HB1, 10X, 1HL, 11X, 1HS, 11X, 2HT2)
      WRITE(6,1130) B1,L,S,T2
 1130 FORMAT(14X, 4F12.4, /)
      WRITE(6,1140)
 1140 FORMAT(20X, 2HPL, 9X, 3HPES, 10X, 2HEL, 10X, 2HER)
      WRITE(6,1150) PL, PES, EL, ER
 1150 FORMAT(14X, 4F12.4, /)
      WRITE(6,1160)
 1160 FORMAT(20X, 2HK3, 10X, 2HK4, 10X, 2HRA, 10X, 2HXA)
      WRITE(6,1170) K3, K4, RA, XA
 1170 FORMAT(14X, 4F12.4, /)
      WRITE(6,1180)
 1180 FORMAT(20X, 2HK5, 10X, 2HK2)
      WRITE(6,1190) K5, K2
 1190 FORMAT(14X, 2F12.4, /////)
      WRITE(6,1200)
 1200 FORMAT(1X, 10HOUTPUT ===, 3X, 1HF, 11X, 2HPT, 10X, 2HPL,
     +10X, 2HPS, 9X, 3HETA)
      WRITE(6,1210) F, PT, PL, PS, ETA
 1210 FORMAT(8X, 5F12.4, /)
      WRITE(6,1220)
 1220 FORMAT(14X, 1HN, 10X, 3HKVA, 10X, 2HIL, 10X, 2HEC, 10X, 2HPA)
      WRITE(6,1230) N, KVA, IL, EC, PA
 1230 FORMAT(8X, 5F12.4, /)
      WRITE(6,1240)
 1240 FORMAT(14X, 1HA, 11X, 1HB, 11X, 1HC, 11X, 1HD, 11X, 1HE)
      WRITE(6,1250) A, B, C, D, E
 1250 FORMAT(8X, 5F12.7, /)
      WRITE(6,1260)
 1260 FORMAT(14X, 2HD1, 8X, 4HB1/L, 10X,2HK5, 10X, 2HK7, 10X, 1HZ)
      WRITE(6,1270) D1, B1DL, K5, K7, Z
 1270 FORMAT(8X,5F12.4, /)
      WRITE(6,1280)
 1280 FORMAT(14X, 2HD2, 7X, 5HT2/D2, 7X, 5HT2/A0, 7X, 5HA0/D2,
     +10X, 2HK2)
      WRITE(6,1290) D2, TOD, TOA, AOD, K2
 1290 FORMAT(8X, 5F12.4, ///)
C  PRINT MESSAGE IF T2/D2 = 0.
      IF(TOD.EQ.C.) WRITE(6,1300)
 1300 FORMAT(27X, 20HTHE LOAD IS MAGNETIC)
 9999 STOP
      END
```

Fig. D.5. (Continued)

Table D.4. Output data for program FALCON

Variable designation	Variable	Units
F	Frequency	kHz
PT	Total power in line, coil, and load	kW
PL	Power into coil and load	kW
PS	Power into load only	kW
ETA	Coil-to-load power conversion efficiency	%
N	Number of turns in coil	. . .
KVA	Value of tuning capacitor	kVA rating
IL	Coil current	A
EC	Capacitor voltage	V
PA	Power loss in transmission line	W
A	Coil resistance RP divided by N^2	Ω
B	Reflected load resistance RS divided by N^2	Ω
C	Coil reactance XP divided by N^2	Ω
D	Reflected load reactance XS divided by N^2	Ω
E	Air-gap reactance XO divided by N^2	Ω
D1	Penetration depth in coil material	in.
B1/L	Diameter-to-length ratio of coil	. . .
K5	Nagaoka's constant	. . .
K7	Magnetic load-resistance coefficient	. . .
Z	$\{(A + B)/[(A + B)^2 + (C + D + E)^2]\}^{0.5}$	. . .
D2	Penetration depth in load	in.
T2/D2	Ratio of load wall thickness to load penetration depth	. . .
T2/AO	Ratio of load wall thickness to load diameter	. . .
AO/D2	Ratio of load diameter to load penetration depth	. . .
K2	Load-resistance coefficient	. . .

Table D.5. Coil-design calculations for solid nonmagnetic stainless steel cylinder

Input

F 10.0000000	S 0.8000	K3 . . . . 0.0000
RHO1 . . . . 0.0000007	T2 . . . . 1.5000	K4 . . . . 0.0000
RHO2 . . . . 0.0000050	PL 100.0000	RA 0.0000
AO 3.0000000	PES 90.0000	XA 0.0000
B1 . . . . 4.0000	EL 440.0000	K5 . . . . 0.9300
L 24.0000	ER 440.0000	K2 . . . . 0.9700

Output

F 10.0000	PA 0.0000	K5 . . . . 0.9300
PT 100.0000	A 0.0000154	K7 . . . . −I
PL 100.0000	B 0.0000253	Z 13.0969
PS 62.2077	C 0.0000153	D2 . . . . 0.0707
ETA 0.6221	D 0.0000261	T2/D2 . . . . . 21.2285
N 18.2230	E 0.0004441	T2/AO 0.5000
KVA 1192.4110	D1 . . . . 0.0291	AO/D2 . . . . 42.4570
IL 2719.5383	B1/L 0.1667	K2 . . . . 0.9700
EC 440.0000		

Table D.6. Coil-design calculations for solid magnetic steel cylinder

Input

F 10.0000000	S 0.8000	K3 . . . . 0.0000
RHO1 . . . . 0.0000007	T2 . . . . 0.0000	K4 . . . . 0.0000
RHO2 . . . . 0.0000050	PL 100.0000	RA 0.0000
AO 3.0000000	PES 90.0000	XA 0.0000
B1 . . . . 4.0000	EL 440.0000	K5 . . . . 0.9300
L 24.0000	ER 440.0000	K2 . . . . 0.9700

Output

F 10.0000	PA 0.0000	K5 . . . . 0.9300
PT 100.0000	A 0.0000154	K7 . . . . 0.0158
PL 100.0000	B 0.0005843	Z 23.7421
PS 97.4341	C 0.0000153	D2 . . . . 0.0707
ETA 0.9743	D 0.0003798	T2/D2 . . . . . 0.0000
N 33.0348	E 0.0004441	T2/AO 0.0000
KVA 139.9376	D1 . . . . 0.0291	AO/D2 . . . . 42.4570
IL 390.8994	B1/L 0.1667	K2 . . . . 0.9700
EC 440.0000		

Table D.7. Coil-design calculations for thin-wall titanium cylinder

Input

F 10.0000000	S 0.8000	K3 . . . . 0.9650
RHO1 . . . . 0.0000007	T2 . . . . 0.2500	K4 . . . . 0.9600
RHO2 . . . . 0.0000210	PL 100.0000	RA 0.0000
AO 2.7500000	PES 90.0000	XA 0.0000
B1 . . . . 3.7500	EL 800.0000	K5 . . . . 0.5533
L 2.2500	ER 800.0000	K2 . . . . 0.9450

Output

F 10.0000	PA 0.0000	K5 . . . . 0.5533
PT 100.0000	A 0.0001540	K7 . . . . −I
PL 100.0000	B 0.0003076	Z 5.2192
PS 66.6445	C 0.0001528	D2 . . . . 0.1448
ETA 0.6664	D 0.0003366	T2/D2 . . . . . 1.7264
N 13.2038	E 0.0036009	T2/AO 0.0909
KVA 886.2114	D1 . . . . 0.0291	AO/D2 . . . . 18.9905
IL 1114.7944	B1/L 1.6667	K2 . . . . 0.9450
EC 800.0000		

Table D.8.   Coil-design calculations for very-thin-wall molybdenum cylinder

Input

F 10.0000000	S 0.7000	K3 ...... 0.0000
RHO1 .... 0.0000007	T2 .... 0.0300	K4 .... 0.0000
RHO2 .... 0.0000200	PL 100.0000	RA 0.0000
AO.......... 4.0600000	PES 90.0000	XA 0.0000
B1 .... 6.0000	EL 800.0000	K5 .... 0.6800
L 6.0000	ER 800.0000	K2 ...... 0.9650

Output

F 10.0000	PA 0.0000	K5 .... 0.6800
PT 100.0000	A 0.0000985	K7 .... − I
PL 100.0000	B 0.0008652	Z 6.5367
PS 89.7791	C 0.0000980	D2 .... 0.1413
ETA...... 0.8978	D 0.0003292	T2/D2 ..... 0.2123
N 16.5367	E 0.0042230	T2/AO 0.0074
KVA 482.5538	D1 ....0.0311	AO/D2 .... 28.7292
IL.........616.0080	B1/L 1.0000	K2 .... 0.9650
EC 800.0000		

References

1. *Ferrous Metallurgical Design,* by J. H. Hollomon and L. D. Jaffe: Wiley, New York, 1947, p 155
2. The Philosophy of Induction Heat Treatment: in *Proc. International Conference on Induction Heating and Melting,* Liège, Belgium, Oct 2–6, 1978
3. *Principles of Heat Treatment,* by M. A. Grossmann and E. C. Bain: ASM, Metals Park, OH, 1964
4. *Steel and Its Heat Treatment,* by D. K. Bullens: Wiley, New York, 1948
5. *Steel and Its Heat Treatment,* by K. -E. Thelning: Butterworths, London, 1975
6. *Heat Treatment of Ferrous Alloys,* by C. R. Brooks: Hemisphere Publishing Corp., Washington, 1979
7. *Principles of Heat Treatment of Steel,* by G. Krauss: ASM, Metals Park, OH, 1980
8. *Metallography, Structures and Phase Diagrams:* Vol 8 of *ASM Metals Handbook* (8th Ed.), ASM, Metals Park, OH, 1973, p 275
9. *Alloying Elements in Steel,* by E. C. Bain and H. W. Paxton: ASM, Metals Park, OH, 1966, p 104
10. Ref. 9, p 112
11. G. A. Roberts and R. F. Mehl: *Trans. ASM,* Vol 31, 1943, p 613
12. U. R. Lenel: *Scripta Met.,* Vol 17, 1983, p 471
13. *Atlas of Isothermal Transformation and Cooling Transformation Diagrams:* ASM, Metals Park, OH, 1977, p 28
14. A. R. Troiano and A. B. Greninger: *Metal Progress,* Vol 50, 1946, p 303
15. Ref 13, p 376
16. Ref 13, p 9
17. Ref 13, p 181
18. *The Making, Shaping, and Treating of Steel,* by H. E. McGannon: U.S. Steel, Pittsburgh, 1971
19. *Metals Handbook* (1948 Ed.): ASM, Metals Park, OH, 1948
20. Ref 6, p 54
21. C. F. Jatczak: *Metal Progress,* Vol 100, No. 3, 1971, p 60
22. Ref 3, p 99
23. Ref 3, p 149
24. C. H. White and R. W. K. Honeycombe: *J. Iron Steel Inst.,* Vol 197, 1961, p 21
25. "Fundamental Aspects of Molybdenum in Transformation of Steel," by E. C. Rollason: Climax Molybdenum Co., London
26. Ref 4, p 472
27. J. H. Hollomon and L. D. Jaffe: *Trans. AIME,* Vol 162, 1945, p 223

28. R. A. Grange and R. W. Baughman: *Trans. ASM,* Vol 48, 1956, p 165

29. J. L. Burns, T. L. Moore, and R. S. Archer: *Trans. ASM,* Vol 26, 1938, p 1

30. W. P. Sykes and Z. Jeffries: *Trans. ASST,* Vol 12, 1927, p 871

31. J. M. Hodge and M. A. Orehoski: *Trans. AIME,* Vol 167, 1946, p 267

32. R. A. Grange, C. R. Hribal, and L. F. Porter: *Met. Trans. A,* Vol 8A, 1977, p 1775

33. L. J. Klinger *et al: Trans. ASM,* Vol 46, 1954, p 1557

34. R. F. Kern: *Metal Progress,* Vol 94, No. 5, Nov 1968, p 60

35. D. C. Buffum and L. D. Jaffe: *Trans. ASM,* Vol 43, 1951, p 644

36. L. D. Jaffe and D. C. Buffum: *Trans. AIME,* Vol 209, 1957, p 8

37. T. J. Dolan and C. S. Yen: *Proc. ASTM,* Vol 48, 1948, p 664

38. *Theory and Application of Radio Frequency Heating,* by G. H. Brown, C. N. Hoyler, and R. A. Bierwirth: Van Nostrand, New York, 1947

39. *Basics of Induction Heating,* by C. A. Tudbury: John F. Rider, Inc., New Rochelle, NY, 1960

40. *Induction Heating Handbook,* by J. Davies and P. Simpson: McGraw-Hill, Ltd., London, 1979

41. *Physics,* by D. Halliday and R. Resnick: Wiley, New York, 1966

42. "Induction Heating Advances: Application to 5800 F," by A. F. Leatherman and D. E. Stutz: NASA Report SP-5071, National Aeronautics and Space Administration, Washington, 1969

43. Ref 39, Vol 1

44. H. G. Carlson: *Industrial Heating,* Vol 23, No. 2, Feb 1956, p 250

45. After a Decade of Solid State Power, by W. F. Peschel: in *Proc. 13th Biennial IEEE Heating Conference on Electric Process Heating in Industry,* Louisville, KY, May 1977

46. Ref 39, Vol 2

47. *Industrial High Frequency Electric Power,* by E. May: Wiley, New York, 1950

48. *Induction Heating Practice,* by D. Warburton-Brown: Odhams Press, Ltd., London, 1956

49. TOCCO Bulletin DB-2034: Park-Ohio Industries, Cleveland

50. *Induction Hardening and Tempering,* by T. H. Spencer *et al:* ASM, Metals Park, OH, 1964

51. *High-Frequency Induction Heating,* by F. W. Curtis: McGraw-Hill, New York, 1950

52. J. T. Vaughan and J. W. Williamson: *AIEE Trans.,* Vol 64, 1945, p 587

53. J. T. Vaughan and J. W. Williamson: *AIEE Trans.,* Vol 66, 1947, p 887

54. R. M. Baker: *AIEE Trans.,* Vol 76, Part II, 1957, p 31

55. W. J. Feuerstein and W. K. Smith: *Trans. ASM,* Vol 46, 1954, p 1270

56. D. L. Martin and W. G. Van Note: *Trans. ASM,* Vol 36, 1946, p 210

57. D. L. Martin and F. E. Wiley: *Trans. ASM,* Vol 34, 1945, p 351

58. *Induction Heating,* by N. R. Stansel: McGraw-Hill, New York, 1949

59. *Fundamental Principles and Applications of Induction Heating:* Chapman and Hall, Ltd., London, 1947

60. *Industrial Applications of Induction Heating,* by M. G. Lozinskii: Pergamon Press, London, 1969

61. Ref 40, p 84–86

62. G. H. Ledl: *Metal Progress,* Vol 92, No. 6, Dec 1967, p 74

63. F. H. Reinke and W. H. Gowan: *Heat Treatment of Metals,* Vol 5, No. 2, 1978, p 39

64. D. E. Novorsky: *Heat Treating,* Vol 13, No. 10, Oct 1981, p 39

65. *Metal Treating,* Vol 24, No. 2, April/May 1973, p 10

66. J. W. Poynter: *Trans. ASM,* Vol 36, 1946, p 165

67. J. W. Poynter: *Trans. ASM,* Vol 40, 1946, p 1077

68. S. L. Semiatin, D. E. Stutz, and T. G. Byrer: *J. Heat Treating,* Vol 4, No. 1, 1985, p 39
69. *Transport Phenomena,* by R. B. Bird, W. E. Stewart, and E. N. Lightfoot: Wiley, New York, 1960
70. B. E. Urband: *Industrial Heating,* Vol 50, No. 4, Apr 1983, p 20 (also B. E. Urband, Tubulars Unlimited, Inc., private communication, Nov 1983)
71. S. L. Semiatin, D. E. Stutz, and T. G. Byrer: *J. Heat Treating,* Vol 4, No. 1, 1985, p 47
72. N. V. Zimin and G. F. Golovin: *Met. Sci. Heat Treat.,* No. 7–8, July–Aug 1966, p 577
73. *Induction Heating,* by W. G. Johnson: ASM, Metals Park, OH, 1946, p 59
74. *Residual Stress Measurements,* by H. B. Wishart: ASM, Metals Park, OH, 1952, p 97
75. O. J. Horger: in *Handbook of Experimental Stress Analysis,* edited by M. Hetenyi, Wiley, New York, 1963, p 459
76. E. D. Walker: in *Residual Stress for Designers and Metallurgists,* edited by L. J. Vande Walle, ASM, Metals Park, OH, 1981, p 41
77. A. Rose: *Harterei Techn. Mitt.,* Vol 21, No. 1, 1966, p 1
78. S. L. Case, J. M. Berry, and H. J. Grover: *Trans. ASM,* Vol 44, 1952, p 667
79. H. Fujio *et al: Bull. JSME,* Vol 22, 1979, p 1001
80. "A Study of the Effect of Induction Hardening Variables on the Residual Stresses and Bending Fatigue Strength of Gears," by R. A. Cellitti: Final Report on Contract No. DA-11-022-ORD-3984, International Harvester Co., Chicago, Jan 1964
81. K. Takitani *et al:* in *Mechanical Working and Steel Processing XVI,* edited by R. N. Edmondson, American Institute of Mining, Metallurgical, and Petroleum Engineers, New York, 1978
82. K. Z. Shepelyakovski: *Metall. i Term. Obr. Metallov.,* Vol 11, Nov 1959, p 24 (H. Brutcher Translation No. 4995)
83. *Distortion in Tool Steels,* by B. S. Lement: ASM, Metals Park, OH, 1960
84. C. A. Siebert and C. Upthegrove: *Trans. ASM,* Vol 23, 1935, p 187
85. C. A. Siebert: *Trans. ASM,* Vol 27, 1939, p 752
86. N. Birks and W. Jackson: *J. Iron Steel Inst.,* Vol 208, 1970, p 81
87. A. Nakashima and J. F. Libsch: *Trans. ASM,* Vol 53, 1961, p 753
88. J. F. Libsch, A. E. Powers, and G. Bhat: *Trans. ASM,* Vol 44, 1952, p 1058
89. N. V. Ross: in *Proc. Sixth Biennial IEEE Conference on Electric Heating,* IEEE, New York, 1963, p 29
90. J. B. Wareing: *High Temperature Technology,* Vol 1, No. 3, 1983, p 147
91. S. Zinn: *Heat Treating,* Vol 14, No. 9, 1982, p 28
92. "State of the Art Assessment of Temperature Measurement Techniques," by H. Pattee: unpublished report, Battelle Columbus Laboratories, Columbus, OH, 1984
93. E. T. Cunnie: in *Proc. Twelfth Biennial IEEE Conference on Electric Process Heating in Industry,* IEEE, New York, 1975, p 74
94. J. Hansberry and R. Vanzetti: *Industrial Heating,* Vol 48, No. 5, May 1981, p 6
95. *Industrial Heating,* Vol 51, No. 3, Mar 1984, p 29
96. P. J. Miller: *Heat Treating,* Vol 14, No. 5, May 1982, p 40
97. P. K. Bhargava and K. M. Joseph: *Tool and Alloy Steels,* Vol 13, 1979, p 195
98. C. L. Kirk: *Metal Progress,* Vol 94, No. 1, July 1968, p 68
99. E. Balogh: *Heat Treating,* Vol 14, No. 9, 1982, p 33
100. "Restrained Induction Hardening and Tempering Axle Shafts with a Non-Destructive Audit for Case Depth," by C. F. Ruhl: SME Paper IQ70-291, SME, Dearborn, MI, 1970

101. Ref 48, p 138
102. P. Lavins: *Metal Progress,* Vol 82, No. 2, Feb 1965, p 109
103. R. E. Bisaro: *Metal Progress,* Vol 104, No. 5, Oct 1973, p 82
104. V. V. Lempitskii *et al: Stal in English,* Vol 5, May 1969, p 499
105. R. E. Jennings: *Metal Progress,* Vol 94, No. 1, July 1968, p 71
106. T. Hijikata *et al:* U.S. Patent No. 4,222,799, Sept 16, 1980
107. R. M. Storey: *Metal Progress,* Vol 101, No. 4, Apr 1972, p 95
108. G. L. Satava: U.S. Patent No. 4,142,923, Mar 6, 1979
109. N. J. Nelson: *Heat Treatment '79* (reprint from proceedings of conference held in Birmingham, England, May 1979)
110. *Machinery and Production Engineering,* Vol 130, Jan 19, 1977, p 65
111. R. Waggott *et al: Metals Technology,* Vol 9, Dec 1982, p 493
112. W. Batz: in *Mechanical Working and Steel Processing XIII,* edited by L. Mair, AIME, New York, 1975, p 138
113. B. A. Kuznetsov: *Metall. i Obrab. Metallov,* No. 2, Feb 1958, p 28 (Brutcher translation No. 450)
114. J. F. Libsch and P. Capolongo: *Metal Progress,* Vol 94, No. 1, July 1968, p 75
115. "Measurements of Elastohydrodynamic Film Thickness, Wear, and Tempering Behavior of High Pressure Oxygen Turbopump Bearings," by K. F. Dufrane, D. Hauser, *et al:* Final Report on Contract NASA-34908, Task 112, Battelle Columbus Laboratories, Columbus, OH, Apr 1984
116. C. W. Marschall: *Metal Progress,* Vol 103, No. 5, May 1973, p 88
117. *Heat Treating, Cleaning and Finishing:* Vol 2 of *ASM Metals Handbook* (8th Ed.), ASM, Metals Park, OH, 1964, p 233 and 236
118. *Industrial Furnaces,* 5th Ed., by W. Trinks and M. H. Mawhinney: Wiley, New York, 1961
119. *Advances in Electric Heat Treatment of Metals,* by N. W. Lord, R. P. Ouellette, and P. N. Cheremisinoff: Ann Arbor Science Publishers, Ann Arbor, MI, 1981
120. *Nondestructive Inspection and Quality Control:* Vol 11 of *ASM Metals Handbook* (8th Ed.), ASM, Metals Park, OH, 1976, p 427
121. *Optical Microscopy of Carbon Steels,* by L. E. Samuels: ASM, Metals Park, OH, 1980
122. *Atlas of Microstructures of Industrial Alloys,* Vol 7 of *ASM Metals Handbook* (8th Ed.), ASM, Metals Park, OH, 1972
123. Ref 122, p 38
124. Ref 122, p 26 and 32
125. Ref 122, p 50
126. Ref 121, p 316 and 317
127. Ref 121, p 322
128. Ref 121, p 347
129. Ref 121, p 326
130. Ref 121, p 348 and 349
131. Ref 121, p 386–388
132. Ref 121, p 389
133. Ref 122, p 33
134. Ref 122, p 30
135. *Principles of Electricity and Magnetism,* by G. P. Harnwell: McGraw-Hill, New York, 1949
136. Ref 40, p 278 and 279
137. G. Falkenbach (unpublished research), Battelle Columbus Laboratories, Columbus, OH, 1984

Index

NOTE. The symbol (F) or (T) following a page-number reference indicates that information is presented in a figure or a table, respectively.